T0135728

Activity Monitoring and Automatic Alarm Generation in AAL-enabled Homes

Aktivitätsbeobachtung und automatische Alarmgenerierung in Wohnungen mit AAL-Technik

Vom Fachbereich Elektrotechnik und Informationstechnik
der Technischen Universität Kaiserslautern
zur Erlangung des akademischen Grades

Doktor der Ingenieurwissenschaften (Dr.-Ing.)

genehmigte Dissertation

von
Dipl.-Ing. Martin Flöck
geboren in Koblenz

D 386

Eingereicht am: 22. Juni 2010
Tag der mündlichen Prüfung: 4. Oktober 2010
Dekan des Fachbereiches: Prof. Dipl.-Ing. Dr. Gerhard Fohler

Promotionskommission

Vorsitzender: Prof. Dr.-Ing. Steven Liu
                           (Technische Universität Kaiserslautern)

Berichterstattende: Prof. Dr.-Ing. habil. Lothar Litz
                           (Technische Universität Kaiserslautern)

                           Prof. Dr. Dr. h.c. Dieter Rombach
                           (Technische Universität Kaiserslautern/ Fraunhofer IESE)

Acknowledgements

This doctoral thesis is the result of my research work conducted at the Institute of Automatic Control at the University at Kaiserslautern from 2007 to 2010. I owe deep gratitude to numerous persons who continuously and patiently supported and encouraged me during this time.

First of all, I would like to thank my doctoral advisor, Professor LOTHAR LITZ, for giving me the opportunity to work and do research in the new and rapidly emerging field of Ambient Assisted Living (AAL). I am grateful for both his invaluable advice and guidance as well as the trust he placed in my work, giving me the freedom necessary for independent research. Without his ongoing support, it would not have been possible to pursue this work to a successful conclusion. Needless to say that I would also like to express my sincere gratitude to the other members of the board of examiners: Professor DIETER ROMBACH (University of Kaiserslautern and Fraunhofer IESE) and the chairman, Professor STEVEN LIU (University of Kaiserslautern).

During my time at the institute, I spent many an hour with my colleagues – both at uni and beyond. I always enjoyed the pleasant and cheerful atmosphere with no room for coercion or quickly taking offence. Instead, our daily routine was characterised by helpfulness, straightforwardness, and the readiness to invest time and energy into pondering about each other's little problems. This certainly does not only hold true for my fellow mates also pursuing their doctorates but also for the permanent employees – without the help and support from our secretary MONIKA KUNZ (who knows all about the unwritten laws...) and our technicians SWEN BECKER, MANFRED ENGEL, and THOMAS JANZ (whose mechanical dexterity is of indispensable value when skilled manual work is to be done), times would have been much tougher here and there.

The foundations of the AAL research at the Institute of Automatic Control, however, were laid by my former colleague MELANIE GROSS. Even though our common affiliation to the institute overlapped only for a few months, I am very thankful for the preparation work carried out by her to make the AAL project become a reality and eventually a success. In addition, I am much obliged to Professor ANNETTE SPELLERBERG, head of the Institute of Urban Sociology at the Faculty of Architecture, Regional and Environmental Planning, and Civil Engineering of the University of Kaiserslautern, and her research group for valuable impulses concerning sociological and ethical issues of our joint AAL pilot project.

All students who contributed to this work certainly deserve my sincere thanks as well, be it that they dealt with the various subproblems of this work as parts of their research papers and diploma or master's theses or as student assistants: MATTHIAS BRINKMANN, NELIA SCHNEIDER, and EMMANUEL TEGUETIO for programming proof-of-concept applications, ASTY NOUKEU and BERND SCHULZE for developing new approaches towards AAL data analysis, VANESSA ROMERO for the graphical artwork for the user interface, BEN BRENDEL for implementing fundamental functionalities of an AAL content management system, and – last, but not least – STEFAN SCHNEIDER for his long-time and reliable assistance throughout the entire time of the project.

Moreover, I am especially indebted to THOMAS GABRIEL and MATTHIAS ROTH for proof-reading the draft versions of this work. Thank you for giving me valuable feedback on the manuscript and assisting me to weed out errata and inconsistencies.

This work was partially funded by the *Ministry of Treasury* of Rhineland-Palatinate (*Ministerium der Finanzen*) and *Stiftung Rheinland-Pfalz für Innovation*, respectively. Additionally, several companies supported the project by means of financial contributions or contributions in kind: *Gemeinnützige Baugesellschaft Kaiserslautern AG*, *Gemeinnützige Baugenossenschaft Speyer eG*, *Gemeindliche Siedlungs-Gesellschaft Neuwied mbh*, and *Wohnbau Mainz GmbH*, partners in the housing sector, *Mobotix AG*, manufacturer of IP cameras, and *Thermokon*, provider of wireless sensor technology. With sincerity, I express my appreciation to all of them. Moreover, I am grateful to *CIBEK GmbH* for providing expert knowledge on installing and configuring home automation components in the Kaiserslautern pilot project.

Finally, and not the least, I wish to convey my deepest gratitude to my wife ULRIKE and my parents. Throughout the time I spent on this work, they showed great understanding, patience, and lenience when there was work in abundance. At the same time, their continuous love and encouragement were an inestimable help for successfully finishing this thesis.

Kaiserslautern, October 2010   MARTIN FLÖCK

Abstract

In this work, novel contributions towards the emerging field of *Ambient Assisted Living* (AAL) are introduced. AAL is a concept envisioned in the early 2000s by the European Commission, aiming at supporting specifically senior people by means of technology and thus helping them to lead independent and self-determined lives in their accustomed surroundings as long as possible. Modern home automation technology is believed to be the key to providing various services in the fields of health, safety, comfort, and communication. In the framework of this thesis, health monitoring aspects are of particular interest. Inactivity monitoring is a very promising approach thereto since it allows the detection of potential health threats or cases of emergency without being overly privacy intrusive. Deriving condensed and dependable inactivity profiles representing typical user behaviour is a pivotal prerequisite for automatic emergency monitoring. Several methodologies for computing such patterns are introduced. Based on those inactivity profiles, various alarming criteria (i.e., permissible inactivity thresholds) are utilised to trigger alarms automatically if the users' inactivity levels exceed individual, user-dependant limits. Since false alarms are inevitable in automatic alarming systems, a procedure of handling them is introduced as well. Finally, the real-world application of the devised AAL system is illustrated.

In dieser Arbeit werden neue Ansätze im Bereich Ambient Assisted Living *(AAL) aufgezeigt. Bei AAL handelt es sich um ein Konzept, das um das Jahr 2000 von der Europäischen Union formuliert wurde. Das Ziel ist, speziell ältere Menschen mit technischen Mitteln zu unterstützen und ihnen so ein unabhängiges und selbstbestimmtes Leben in vertrauter Umgebung so lange wie möglich zu erlauben. Moderne Hausautomatisierungstechnik wird dabei als Schlüssel für das Bereitstellen unterschiedlichster Funktionen in den Bereichen Gesundheit, Sicherheit, Komfort und Kommunikation angesehen. Im Rahmen dieser Dissertation gilt Gesundheitsaspekten besonderes Interesse. Inaktivitätsmonitoring ist in diesem Zusammenhang ein vielversprechender Ansatz, da es die Erkennung möglicher Gesundheitsgefahren oder Notfälle ohne massiven Eingriff in die Privatsphäre der Nutzer erlaubt. Die Generierung komprimierter, aussagekräftiger Inaktivitätsprofile, die das typische Nutzerverhalten abbilden, ist eine wichtige Voraussetzung für die automatisierte Notfallerkennung. Mehrere Verfahren für die Erstellung solcher Profile werden vorgestellt. Auf der Basis dieser Inaktivitätsprofile lösen verschiedene Alarmkriterien, d.h. zulässige Inaktivitätsgrenzen, automatische Alarme aus, wenn die Inaktivitätsdauern der Nutzer individuelle, personenspezifische Grenzwerte überschreiten. Da Fehlalarme in automatischen Alarmierungssystemen unvermeidbar sind, wird ferner ein Procedere zum Umgang damit vorgestellt. Abschließend wird die praktische Umsetzung der beschriebenen Verfahren anhand zweier Projekte beschrieben.*

Contents

1.   Introduction

1.1.   Motivation

Life expectancy has been steadily increasing in most industrialised countries such as the member states of the European Union (EU), North America, or Japan, over the last decades. The objective of this work is thus providing *Ambient Assistive Living* (AAL) technologies, based on standard home automation components, to mitigate the impact of this demographic change on seniors and society in its entirety. In a nutshell, AAL comprises all technological approaches from which seniors can benefit in their everyday life and that help them to maintain their independence and autonomy as well as to stay involved in social activities in their neighbourhood. It is believed by the author that monitoring activity and inactivity as well as automatically raising alarms if unexpected behavioural patterns are found in order to safeguard their well-being is one of the keys to enabling seniors to lead such independent and self-determined lives in their accustomed surroundings. The following considerations underscore the need for novel technological approaches and strategies to cope with the demographic change.

From 1996 to 2007, the average life expectancy in Germany rose by 3.8 years in men and 2.6 years in women, respectively [Eurostat, 2009b]. At the same time, many countries will face severe changes in their population size. Within the EU the total population size is predicted to remain at a constant number of individuals over the period from now on to the year 2060. Broken down to the individual countries, however, the situation looks fundamentally different. Numerous countries will see substantial increases in their population (e.g., Ireland (+2.14 m / +46.4%), the United Kingdom (+14.7 m / +23.7%), or France (+9.21 m / +14.7%)). Since the EU's predicted total population is not to change, other EU states' population sizes will have to decline accordingly. Examples for the respective downward tendency are Germany (–11.38 m / –13.9%), Poland (–6.95 m / –18.2%), or Romania (–4.41 m / –20.7%). On closer inspection, the data provided by Eurostat suggests that a migration from the East to the West and the North of Europe will set in that might help to explain the above figures [Eurostat, 2009a].

Moreover, there is reason to assume that not only will the life expectancy increase but also will the old-age-dependency ratio shift towards higher values, meaning that there will be substantially more people aged 65+ compared to the labour force of working age. The predicted figures for the EU indicate that this trend will take effect no matter how the population size in general will develop. Considering the above example EU states, e.g., the old-age-dependency ratio of Romania will increase from 21.3% to 65.3%, whereas it will rise from 16.7% to 43.6% in Ireland [Eurostat, 2008].

This imminent development poses various challenges as all members of society are affected thereby. Societies need to take up these challenges now in order to maintain a high standard of living for the affected age groups and to guarantee –if needed– an adequate level of care for those who need home or institutionalised care. Health care systems will be particularly impacted by the ageing population because of the increasing demand of care services by the elderly. This increasing demand cannot be satisfied by the current workforce employed by health care providers. In addition, the availability of modern medical technology as well as better educated, wealthier, and thus more demanding elderly also contribute to the increasing work load of care staff. Projections for the number of health care workers needed to maintain the current level of care were released by the World Health Organization as background for the World Health Report 2006. It is estimated that in 2050 13.8% of the active work force in Western Europe will have to be health workers (+5.4 percentage points) and 14.2% in the USA/Canada (+4 pp), respectively [Matthews et al., 2006].

From the author's point of view, modern technology can help to mitigate the implications of this development outlined above. Since both financial means and manpower of care providers are limited resources [Fugger et al., 2007], postponing the point in time when seniors start needing informal or formal care is an important starting point. This is where modern home automation technology comes into play. By designing smart environments capable of keeping track of the health status of the resident, commonly referred to as Ambient Assisted Living (AAL) environments, seniors are enabled to remain in their accustomed homes as long as possible. Monitoring the activity and inactivity levels of the residents over extended periods of time allows generating profiles that represent typical user behaviour and –based thereupon– detecting anomalies in the daily routine of a resident that may be indicative of a possible health problem. If such an anomaly is found, automatic alarms can be raised to facilitate administering help and taking all necessary steps to ensure the user's health and well-being. Ultimately, being able to stay at the family home independently rather than requiring formal or informal care and postponing relocating to a residential home will, on the one hand, contribute to quality of life and, on the other hand, be a step towards reducing expenditures of the care providers and thus health care schemes.

Although health monitoring is the most important objective in the AAL approach followed in this work, additional functionalities were implemented, e.g., relating to comfort, safety, and communication, in order to boost the acceptance on the side of the users. Details on this approach will be discussed in the sections below.

1.2.   Novel Contributions and Targets of this Work

Over the course of many decades, sensor and actuator technology has gradually been introduced into private homes. One of the first sensor-based technologies widespread use was made of was the burglar alarm. EDWIN HOLMES was the first to develop burglar alarms for serial production as early as in the eighteen fifties. In 1880, he marketed a device for automatic illumination of a house when an alarm was tripped [McCrie, 1988]. Only in 1969, the photoelectric smoke detector was patented [Smith & House, 1969]. In 1980, the first personal emergency response system (PERS) was introduced by AEG-Telefunken and St. Willehad Hospital in Germany [Paulus & Romanowski, 2009]. This timeline illustrates the basic level at which all automation technology for safety and monitoring purposes developed until then. Up to that point, most devices were isolated solutions not able to communicate with each other or having intrinsic intelligence. Eventually, the predecessors of today's KNX home automation standard, namely EIB, were introduced in the early nineteen nineties [KNX, 2006].

Meanwhile, the state of the art of personal medical alarms has indeed evolved as well, but they are still far from being smart in the sense of ambient intelligence (AmI) [Lindwer et al., 2003, van der Poel et al., 2004] or AAL, a subset of AmI (see 2.2 for the various approaches towards smart environments). Some of the basic principles of the AmI vision are that technology should be invisible, embedded in the user's surroundings, adaptive to the user's needs, and acting autonomously. This is where this project starts from. A catalogue of requirements with regard to AAL was specified. It is assumed that adhering to these requirements will lead to a successful AAL implementation within the work conducted in this thesis:

+ **Detection of medical conditions and emergencies:**
The *overall* goal of the developed AAL solution is to provide the user with assistive technology. The main task of this technology is safeguarding the health and safety of the user who relies on the capability of the system to detect anomalies in their daily routine.

+ **False alarm handling and multi-level alarms:**
False alarms play an important role when dealing with personal emergency response systems. They cannot be avoided; quite the contrary, they are an integral part of any monitoring system or personal emergency response system. False alarms can even be beneficial due to the training effect on the users – an occasional false alarm will make sure that the user knows how to handle them and that he gains confidence in the system as well as the operators at the emergency response call centre [Adam, 2009].

   In order to reduce the number of false alarms being forwarded to emergency response services, a multi-level alarm scheme will be applied. In case of an alarm, the user will be alerted directly. This way, the user can acknowledge the (false) alarm and thus prevent that anyone outside the flat or the house will be notified. Only if the alarm is not acknowledged, it will be forwarded to the emergency response service. This concept will be discussed in detail in section 7.2.

+ **High acceptance among future users:**
Any solutions developed aiming to be AAL technology are made for *people*. Thus, meeting the users' needs and wants is of crucial importance. User integration and participation are believed to be key factors for ensuring high acceptance.

+ **User integration and participation:**
Part of the endeavour to achieve very high user acceptance is the design process followed in this project. Instead of using the classic software development method which is referred to as the *waterfall model* (Fig. 1.1), an iterative development process is to be implemented (Fig. 1.2) [Royce, 1987].

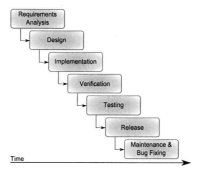

Fig. 1.1:   Classic waterfall method for software development [Royce, 1987]

The disadvantage of the waterfall model is that the development process is rather slow and that not all flaws and shortcomings of the developed AAL solution can be detected before the product is actually deployed. Any bug-fixes and corrections have to be done after the final product has been shipped to the user.

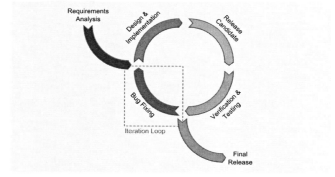

Fig. 1.2:   Iterative development process [Rombach & Ulery, 1989, Boehm et al., 2005]

In contrast to that, the iterative development process as shown in Fig. 1.2 allows user participation because all release candidates are deployed to the users. Hence, immediate feedback about bugs and malfunctions can be collected in order to improve the AAL solution. In addition, user involvement guarantees that the developed AAL solution does indeed meet the users' expectations and needs as identified during the requirements analysis or the verification steps. This procedure is commonly referred to as a human centred approach [Borges et al., 2008]. A detailed explanation of the particular user feedback approach followed in this project will be given in section 3.3.

+   **Ease of use:**

Since the prospective users are seniors, particular attention has to be directed towards usability. The entire system has to be designed in such a way that it is firmly based on a well thought-out usability concept so that it can easily be grasped by senior users. This includes that –if present in an AAL environment– graphical user interfaces (GUIs) must be specifically designed according to the needs and impairments of the target group. In the current work, all flats are equipped with a central control unit called the *P*ersonal *A*ssistive *U*nit for *L*iving (PAUL) featuring a GUI. Details about the particular design of PAUL will be discussed in section 8.2.3.

+ **Unobtrusiveness:**
In order to respect the privacy of the users and to guarantee their right to informational self-determination, the AAL system should be as unobtrusive as possible. Hence, just enough data is to be collected and processed to attain the overall objectives that are set forth in this section.

+ **High reliability:**
The technical basis for collecting activity signals and providing comfort and safety functionalities (e.g., sensors, actuators, and home automation bus systems) should be commercially available and technically mature. Thus, off-the-shelf components were chosen.

+ **Flexibility:**
The AAL solution to be developed should be as flexible as possible. This means that it should be applicable both in newly built houses as well as in the existing housing stock which is the prevalent field of application as most seniors live therein. Moreover, various bus systems shall be supported since home automation technology advances rapidly and dependence on a specific bus system should be avoided.

+ **Commercial viability:**
Last but not least, commercial viability is also of crucial importance. Only if marketability is guaranteed which in turn a business model can be built upon, as many prospective users as possible will benefit from the development of AAL technology.

To boost user acceptance and give prospective users as well as housing societies additional incentives to buy and install the new AAL technology, additional functionalities will also be integrated seamlessly into the concept. It is envisioned to upgrade the AAL environment to a level that it provides safety, comfort, communication, and health monitoring in a truly ambient manner.

1.3.   Organisation of the Thesis

In this subsection, a brief overview of the organisation of the thesis will be given.

Section 1 is the introductory chapter in which the motivation for this work and the novel contributions to the state-of-the-art are explained.

Section 2 addresses general aspects of Ambient Assistant Living. First of all, the history of AAL and its predecessors –Ubiquitous Computing and Ambient Intelligence– will be briefly outlined. Subsequently, the state-of-the-art of current AAL research and related works will be summarised. The section will conclude with fundamental reflections on ethical and privacy implications raised by AAL research in general.

The development process followed in this thesis will be the topic of section 3. First of all, the analysis of the users' requirements is discussed. However, as outlined in the previous subsection, user involvement is pivotal in order to shorten development cycles and to make sure that the users' needs and wishes are indeed fulfilled. Hence, user feedback forms an essential part of the development process. Other than classic approaches based on a linear development process, an iterative development strategy is applied. New functionalities are only introduced step by step to allow for the gathering of user reports on utility and stability thereof.

In section 4, the process of the sensor data collection, pre-processing, interpretation, and storing is elucidated. During the process of analysing the users' inactivity profiles, the sensor raw data is transformed and condensed several times, ultimately yielding the so-called tertiary data.

Section 5 addresses the fundamental difference between monitoring activity and inactivity as well as the respective advantages and disadvantages of these two approaches. In the course of this work, it turned out that the inactivity approach is superior to activity analysis. More-over, the different natures of singular events and continuous activity –as captured by motion detectors– will be discussed. Finally, the actual inactivity patterns used in this work will be introduced.

Section 6 is concerned with alarm generation. At first, multi-day inactivity patterns estab-lished by merging inactivity data sets of multiple days will be introduced. Subsequently, the concept of alarm thresholds will be discussed. Alarm thresholds are established on the basis of the inactivity pattern of a person monitored. If the current inactivity in the flat or house exceeds the threshold, an alarm will be triggered. Advanced monitoring techniques that allow improving the sensitivity of the basic inactivity monitoring will be elucidated as well. Taking additional knowledge, e.g., about changes in the behavioural patterns due to changing sea-sons, holidays, or working shifts, into account contributes to the refinement of the data analy-sis with regard to generating alarms. As a result, otherwise inexplicable "anomalies" can be put down to peculiarities of the normal behaviour of the person being monitored.

In section 7, the multi-level alarm scheme put in place to avoid excessively large numbers of false alarms will be explained. When an alarm is raised, the user will always be given the chance to acknowledge it in order to cancel it. Only if the alarm is not acknowledged, it will be forwarded to an emergency response service. Moreover, the user will be able to set the *mean time between false alarms* (MTFA) and thus determine the sensitivity of the AAL sys-tem. If a large MTFA is chosen, only few false alarms will be tolerated so that the response time of the system will be comparatively long and vice versa. The effect of the MTFA on the applicability of different alarm thresholds will also be addressed.

In section 8, actual practical implementations of the developed AAL system will be illus-trated using the examples of two real-world installations. First, the graphical user interface used in these installations will be introduced. The special requirements concerning the devel-opment of a GUI particularly suitable for seniors will be explained. Additional functionalities in the fields of comfort, communication, and safety that were added as incentives for the user to accept and use the device in their daily life will be presented. Moreover, shortcomings of the used home automation hardware will be discussed. The section concludes with a brief overview of the results of the socio-scientific research that has been conducted to assess the suitability of the developed AAL solution for the target group, i.e. aged people. Second, the technical details of the two installations are described. The chapter concludes with both an overview of technical challenges encountered in the course of the practical implementation and a summary.

In section 9, the contents of this thesis will be summarised. Section 10 is composed of the bibliography and indexes. Appendices to this work can be found in section 11.

On a side note, the terms *inactivity pattern* and *inactivity graph* are used synonymously in the following. Moreover, it needs to be noted that in every instance in which real personal data of a tenant are displayed, neither the tenant nor the month in which it had been collected will be revealed in order to protect the privacy of the tenants. Instead, summer months are labelled Sn and winter months Wn.

2.   Ambient Assisted Living (AAL)

2.1.   Introduction

The term AAL first appeared in the European Commission's terminology in the early 2000s. In 2008, the first AAL Joint Programme Call was launched by the national member organisations in the individual EU countries [EU, 2008]. Numerous definitions of Ambient Assisted Living (AAL) have since been published. A common basis, however, can be found when comparing several definitions in the following.

Supporting the elderly by means of modern technology is a frequently recurring theme in all of the definitions. Support can be provided in various ways and for various reasons. It is common to all definitions that particularly the elderly shall benefit from AAL technology in their everyday life in order to sustain their independence and autonomy and to help them cope with situations in which they would require assistance otherwise. As a result, senior people shall be enabled to maintain their self-determination and to actively participate in social activities as long as possible.

The domains assistive technologies (AT) shall be employed in are comfort, safety, and health. Summarising the cited literature, the visions expressed include easy-to-use AT that is context-aware, adaptive, and sensitive to the users' needs. AT is envisioned to compensate for age-specific deficiencies and thus give the users the freedom to stay in their accustomed homes as long as possible [Bechtold & Sotoudeh, 2009, Becks et al., 2007, EU, 2006, Floeck & Litz, 2007, Floeck & Litz, 2008b, von Foerster, 2008, Georgieff, 2008].

Although considerable common ground can be found between the different approaches described in literature, the difference is in the details. One of the major dissimilarities is related to the degree of ambience (i.e., the inconspicuousness of the technical components) that is to be achieved. At one end of the spectrum, there are approaches involving extensive data acquisition using numerous sensors, several of them being worn on the body (e.g., [Lukowicz et al., 2004, Noury et al., 2000, Winters & Wang, 2003]). Particularly the projects targeting treatment of and (tele-)care for patients with chronic illnesses may supplement the body-worn sensor array with sensors for specific pieces of information (e.g., bed sensors) or specialised diagnostic units for determining medical parameters, e.g., blood test results (e.g., [Kaufman et al., 2003, Kung et al., 2007, Leijdekkers et al., 2007, Nambu et al., 2000]). The measured vital signs may be transmitted to medical professionals for further interpretation.

Depending on the particular realisation, however, concepts involving body-worn sensors or the acquisition of vital signs can be rather intrusive. Hence, at the other end of the spectrum, several approaches exist that merely rely on ambient standard sensors seamlessly integrated into the surroundings in order to mitigate ethical concerns and improve user acceptance. These ambient sensors are used for collecting generic information about the activity of the user, thus allowing the analysis of activity or even activities of daily living (ADL) (e.g., [Barger et al., 2005, Barnes et al., 1998, Beckmann et al., 2004, Floeck & Litz, 2008a, Floeck

& Litz, 2009, Glascock et al., 2007]). By merging and interpreting these different kinds of information, knowledge about the daily routine, health, and potential emergencies can be obtained. In section 2.3, an overview of selected AAL projects worldwide will be given.

Another important difference in the objectives being pursued by AAL researchers is the kind of anomalies or emergencies that are to be detected (see Table 2.1). Emergency prediction on the one hand and emergency detection on the other hand need to be distinguished [Nehmer et al., 2006]. Emergency prevention involves diagnosing the onset or exacerbation of chronic illnesses such as Alzheimer's or dementia, i.e., processes taking days, weeks, or even months to become apparent and require attention. In contrast, emergency detection means identifying sudden emergencies within a much shorter time span of minutes to hours. Sudden emergencies, however, cannot be regarded as a homogeneous class of incidents. Some of them need to be acted upon immediately (e.g., heart attacks or strokes) whereas others are less critical (e.g., falls).

Table 2.1: Severity levels of domestic emergencies or medical conditions for AAL

Severity level	Description	Life-threatening	Examples	Action required
A high	Sudden, genuine emergency	Yes	Heart attack Stroke	Immediately (within few minutes)
B medium	Sudden, genuine emergency	No	Fall Unconsciousness	As soon as possible (within minutes to few hours)
C low	Onset or exacerbation of a chronic illness	No	Hypertension Dementia Decrease of activity	Subject should see a doctor (within days to weeks)

The likelihood of an incidence of any of the severity levels listed in Table 2.1, however, greatly depends on the age of the individual being monitored. In geriatric terminology, the so-called elderly are commonly divided into three groups and referred to as the young old (aged 65–74), the old old (aged 75–84), and the oldest old (aged 85+) [Fornara et al., 2001]. In addition to that, a fourth group –aged 50–65– can be defined. They play an increasingly important role as customers possessing purchasing power. However, even this age group is considered inhomogeneous [Kimpeler et al., 2006]. These remarks illustrate the enormous lack of consistence in the needs of the targeted age group and thus their vastly varying needs and expectations. This is to be kept in mind when designing AAL solutions for "the elderly".

2.2.   Artificial Intelligence, Ubiquitous Computing, Ambient Intelligence, and Cyber-Physical Systems

Ambient Assisted Living is a comparatively new term. Numerous technologies, however, can be associated with it, either as technological predecessors or as more advanced visions currently being under development. Artificial Intelligence (AI), Ubiquitous Computing (UC), and Ambient Intelligence (AmI) paved the way for AAL, whereas Cyber–Physical Systems are the next evolutionary step towards truly ambient, autonomous, and intelligent systems.

The idea of Artificial Intelligence had first been discussed by ALAN M. TURING in 1950 ("can a machine think?") [Turing, 1950]. In this groundbreaking paper, the idea of intelligent machines had been introduced. Although AI has undergone rapid development and diversification, particularly since the mid-1980s, the state-of-the-art of AI still has not yet reached genuine intelligence; it can rather be characterised as *advanced computing* [Munakata, 2008]. Today, research on AI is being conducted in a vast array of fields, e.g., uninformed and informed search, knowledge representation, machine learning, or neural networks [Jones, 2007]. All of these play an important role for AAL. It needs to be noted, however, that AI is controversially discussed among scientists and that misconceptions about the nature of AI are widespread [Mira, 2008, Wang, 2007].

In 1991, MARK WEISER presented his vision of Ubiquitous Computing (UC) [Weiser, 1991]. In this seminal paper, the idea of technologies that "weave themselves into the fabric of everyday life until they are indistinguishable from it" is first expressed. Today's computer technology is being criticised for its complex operation that has nothing to do with the tasks it is used for. Thus, it is believed by WEISER that our current computers are only a transitional step towards their disappearance into the background of everyday life. The requirements for the achievement of this stage of invisibility are small, cheap, and low-power nodes, software suitable for UC applications, and wireless networks connecting the nodes. A flat computer screen is envisioned to be the user interface to operate the UC environment.

The next step towards the concept of AAL was the EU's announcement of their Ambient Intelligence (AmI) programme [EU, 2004]. The requirements for AmI applications outlined in this paragraph and those for AAL are largely congruent. The below AmI characteristics complement the specifications of AAL systems given in sections 1.2. The aim of the AmI programme is to provide information services to everyone at any time by means of a vast variety of devices. In order to implement these services, an in-home wireless network, personal area networks for the individual residents and a server holding both the personal and shared data need to be available. Eventually, a world comprising a large number of easy-to-use interfaces, seamlessly embedded into everyday objects, is envisaged. This AmI-enabled environment is to recognise and respond to individuals in an unobtrusive way, thus assisting them in living, working, and relaxing (together). Hence, AmI systems need to be

+ context-sensitive,
+ personalised,
+ able to reconfigure themselves dynamically to unforeseen conditions,
+ able to adapt to varying needs of the user,
+ compatible to neighbouring AmI systems,
+ failsafe, even if parts of the network malfunction, and
+ trustworthy, i.e., safeguarding the privacy, security, and safety of the user.

Cyber-Physical Systems (CPS) are the most recent step forward in the development of ambient, smart, and autonomous computing [Wolf, 2009]. This novel concept was first envisioned by the U.S. National Science Foundation in 2006, aiming at the development and implementation of computer- and information-centric physical and engineered systems. The genuine novelty of this technology is the fusion of computation and physical processes: The fundamental principle of a CPS is that embedded systems monitor and control physical processes and physical processes affect the computations executed by the embedded systems [Lee, 2008]. The intended characteristics of CPS are far beyond what UC or AmI were designed for, namely being highly dependable and capable of advanced performance in computation, communication, and control. There is, however, a significant overlap with the above existing approaches in the projected applications: health care, assisted living, bionics, and wearable devices are expressly listed as potential application domains [NSF, 2006].

2.3.   Related Works

2.3.1.   Preliminary Considerations

Numerous research groups worldwide are working in the field of Ambient Intelligence. For example, the work presented in this paper is embedded into the joint Ambient Intelligence research effort within the *ambient systems "amsys" research centre* at the University of Kaiserslautern [amsys, 2010]. Numerous scientists from various disciplines –electrical engineering, mechanical engineering, computer sciences, and social sciences– are researching several aspects of Ambient Intelligence and Ambient Systems. Amongst others, the Fraunhofer Institute for Experimental Software Engineering is conducting research in the field of identifying activities of daily living (ADLs) which is, however, not focus of the present work [Nehmer, 2009, Nehmer et al., 2006, Kleinberger et al., 2009, Storf et al., 2009]. Moreover, at the University of Kaiserslautern, for example measuring vital signs has been investigated [Ayari & Tielert, 2007] or concepts for the training of professional cyclists have been developed [Le et al., 2007]. The plenitude of different topics Ambient Intelligence research can comprise illustrates that is it crucial to focus on a limited scope to able to achieve significant results. Since AAL is the focus of this work, in the following only research groups working in this field as well will be considered.

However, even when only taking into account AAL research, the various approaches need to be distinguished in several ways, e.g., with regard to the objective (maintaining independence, emergency avoidance and detection, monitoring chronic illnesses, etc.) or the means of collecting data (ambient sensors or sensors worn on the body). Another important criterion is whether medical staff will have access to the data or whether the AAL system is working stand-alone. In the following, a brief overview of selected important approaches the author is aware of will be given. The attempt to give a comprehensive account of all research endeavours would go far beyond the scope of this section. Three main realms for the application of assistive technology have been identified and will be presented below (2.3.2–2.3.4). This section will conclude with a brief overview of the impact of AAL technology on the life of seniors.

2.3.2.   Telecare and monitoring chronic illnesses/vital functions

Medical conditions and vital functions can be observed for multiple reasons. First, telecare solutions can help users to cope with already existing illnesses. Second, vital functions can be monitored in order to assist medical staff in caring for the affected individual. Third, long-term trends may be attempted to be identified to allow early intervention in case of a deteriorating health state.

In [Reyes Alamo et al., 2009], an approach is presented that illustrates how individual users can be supported in handling their illnesses independently. The work aims at integrating and standardising the management of personal medication into the OSGi framework, a technology widely used in a huge variety of applications, building and home automation being one thereof.

Numerous works also tackle the collection of vital signs for evaluation by medical staff. In [Zito et al., 2007], a miniaturised, wearable sensor for the collection of heart and respiratory rates was developed. In [Anliker et al., 2004], a wrist-worn device for the collection of multiple vital signs and their wireless transmission using a cellular link to a medical centre was implemented. These two works try to give (high-risk) patients the greatest possible degree of independence and freedom while at the same time guaranteeing that their vital functions are being monitored continuously. Hence, emergencies can be detected instantly.

Last but not least, one of the paramount goals of telecare and AAL is modelling the users', i.e., patients', profiles that can either hint at existing deficiencies or –when monitored for extended periods of time– the onset of illnesses such as dementia. In [Virone & Sixsmith, 2008], an approach involving an avatar, i.e., a computer-generated assistant, encouraging the users to perform their chores, is presented. At the same time, long-term 24-hour activity patterns (circadian activity rhythm, CAR) are created and analysed. If deviations in the activities of daily living (ADL) are observed, an alarm is generated. Several alarm classes are being distinguished, e.g., too little or too much activity or daytime vs. nighttime trends. As a result, caregivers are given a mean for early identification of possible cognitive declines. A similar approach is followed by [Hadidi & Noury, 2009]. In this paper, motion detectors were used in order to obtain information about various parameters, e.g., diurnal and nocturnal activity, respectively, or the number of activity events observed by individual sensors. The collected profiles were matched with the health record of the patients and yielded promising results regarding the assessment of ADL, thus enabling them to stay in their homes independently.

2.3.3.   Emergency avoidance and detection

In [Nehmer, 2009], a visionary model of human vitality is introduced. Human vitality is divided into three regions, i.e., healthy, non-critical, and critical. Furthermore, this model hypothesises that even with the most advanced technology, the vitality of an individual can only be approximated. Hence, emergencies can only be detected with a certain delay in time. In order to anticipate and detect emergencies as early as possible, vital signs, motion, or ADLs need to be monitored continuously. If the central data evaluation unit believes to have detected any abnormal behaviour, possibly indicating an emergency, it will try to contact the individual through a bidirectional audio interface, prompting the acknowledgement of the emergency. If the emergency is confirmed or no answer is given, further actions will be initiated by the assistance system. In [Kleinberger et al., 2009], the EMERGE project is described in which unobtrusive sensors are used to assess the potentially arising emergency situations. In this approach, a combination of sensors, software components, and expert systems is used for situation recognition and providing appropriate assistance. Short-term emergencies (e.g., falls, critical vital signs) and critical deviations from long-term behavioural patterns are distinguished. Even though considerable detections rates for ADL could be achieved, problematic tasks requiring further research were also identified. Malfunctions of sensors, too little a number of sensors to guarantee that no ADLs slip through, or unexpected behaviour of the user (e.g., acting too fast) are some of the problems encountered and need to be tackled for the presented AAL solution to become fully operational. An approach for the detection of highly specific problematic conditions (e.g., falls, high and low pulse rate, forgotten stove) is described in [Dalal et al., 2005]. Both a self-learning algorithm and a rule-based algorithm based on expert knowledge were compared. Although both algorithms performed quite well under standard conditions (same ground plot of flat, similar behaviour of the users), unacceptably high rates of false alarms were observed when deploying the technology to individual homes with highly variable ground plots and behavioural patterns of the users. Further work needs to be done to overcome these limitation and to make this approach more generic and thus viable under a plenitude of conditions.

One particular field of research, especially targeted at seniors living single, are assistive technologies for fall detection. Although fall detection is no key objective of this work, it is important to keep in mind that considerable effort is devoted to this problem in order to detect this class of emergency and to counteract the implications it can have, i.e., early formal care in nursing homes or even hospitalisation. Various approaches are followed by numerous research groups worldwide. In [Yu et al., 2009], a concept involving a single camera is described, allowing the detection of a fall by discriminating between standing, bending, and lying postures. Another approach is described in [Zhang et al., 2006]. In this work, mobile phones were equipped with accelerometers and advanced signal processing algorithms to detect falls of the person carrying the phone. As a result, falls could be detected effectively without violating the users' privacy by using intrusive sensors. A single camera approach has also been proposed by [Schulze et al., 2009]. In order to detect falls, the outline of a person is continuously tracked. Cease of inactivity or prolonged sojourn times at a specific position are assumed to be indicative of a possible fall.

2.3.4.   Maintaining independence, increasing quality of life

Many AAL research groups strive to enable senior to live in their accustomed homes as long as possible. This aspect of AAL research is particularly important since it contributes both to maintaining independence and thus quality of life and at the same help to reduce expenditures of public health schemes. In [Cook, 2006], it is hypothesised that automated assistance and health monitoring systems can empower seniors to lead independent lives in their own homes. In order to achieve this, an intelligent environment had been designed that is capable of acquiring and applying knowledge about its inhabitants in order to adapt to them and provide comfort, safety, and health. Thus, a large number of various sensors is used to collect information about the ADLs of the inhabitant. This data is subsequently mined for patterns that allow the prediction of upcoming events so that the smart environment can anticipate them and react accordingly to the predicted wishes of the inhabitants. General assistance could successfully be provided, but health-specific assistance has not yet been implemented.

The outcomes of the I.L.S.A. (Independent LifeStyle Assistant) as described in [Haigh et al., 2006] show that designing and implementing AAL environments is not as straightforward as one may be inclined to believe. The I.L.S.A. project had a large scope and yielded many valuable results, but encountered numerous difficulties in the details. To name but a few, the data collection (expensive hardware, trip hazards due to sensor installation, complex data evaluation, resistance of users towards intrusive sensor technology), the user interface (speech recognition and generation complex and likely to be refused), or automated reasoning (recognising abandoned activities, user working concurrently on more than one task) posed considerable challenges. It can be concluded that narrowing down the objectives of future efforts to assess and interpret ADLs at the beginning may be beneficial. More advanced goals should only be pursued when basic functionality is reached.

In [Mynatt et al., 2000], a concept addressing three aspects of AAL is presented. First, crisis recognition is part thereof. It is not only limited to acute health threats (e.g., falls) but also includes detecting and solving problems related to "normal" technical equipment (e.g., failure of a heater in winter). Second, it aims at supporting everyday cognition. Cognitive deficiencies, e.g., forgetting to take medication, shall be compensated for. Third, maintaining bonds with family members and being reminded of them is a key part of this approach. This is believed to be an important factor for long-term health and happiness that cannot easily be quantified in scientific terms.

In summary, it becomes clear that there is a vast array of factors that need to be considered when designing and developing AAL environments and solutions for the elderly. Numerous research projects are currently being conducted, but, however, most of them have not yet reached a state that would allow the deployment to the end user. Further research needs to be carried out to reach viable solutions that do indeed meet the users' needs and wants without being intrusive and that fulfil the expectations that health professionals and family members have.

2.3.5.   Acceptance of monitoring technologies by and their impact on seniors

The Kaiserslautern AAL project described in this work has been evaluated by [Spellerberg et al., 2009]. The assessment revealed that after two months all users were able to easily operate the AAL user interface. The users consistently declared that the graphical user interface is easy to use. 18 out of 19 tenants use PAUL regularly. The most frequently used functionalities are the video entry phone, the remote control of the roller blinds, and Internet access. The overall results are that the inhabitants of Kaiserslautern project are glad to participate in a scientific research project. Moreover, the AAL technology is an important issue to talk and chat about with neighbours, thus improving social bonds and integration as well as the community spirit merely by its existence, let alone the assistance it provides in everyday life. Another pivotal finding of the acceptance evaluations was that the inhabitants of the AAL-enabled block of flats do not perceive the developed AAL solution as intrusive, privacy-violating, or interfering with their self-determination. On the contrary, when prompted for their opinion regarding the collection of activity data for monitoring their health status, only positive feedback was given by the users. It is further assumed that easing everyday life by the provision of safety, entertainment, information, and communication functionalities contributes to an increased quality of life while living in the accustomed homes.

In [Alwan et al., 2006], similar findings obtained in a study conducted with 22 individuals requiring formal care are reported. The technical equipment included a bed sensor, motion detectors, and a stovetop temperature sensor. Only 2 out of 24 persons approached declined to participate in the study because of refusing being monitored. Interviews revealed that the perceived quality of life (QoL) of the participating individuals increased during the first three month of the project, both in terms of a higher average QoL and a significantly reduced variance in the QoL. This study concludes that activity monitoring and sharing the data with healthcare professionals were acceptable to older adults when in return an increased sense of safety and timely intervention in case of an emergency were to be anticipated.

2.4.   Ethical and Privacy Implications

2.4.1.   Ethical Implications

The use of information and communication technologies (ICT) as well as monitoring technology in an AAL context raises a number of ethical questions. Most researchers addressing this issue argue that insufficient attention is directed to ethical concerns. At the same time, others comment that ethical concerns are voiced excessively [Kubitschke et al., 2009]. As a result, the development of innovative AAL solutions for a wide range of applications may be stifled [Mahoney et al., 2007]. Within the scope of this work, this area of conflict cannot be discussed exhaustively, but this section will give a brief account of which questions need to be taken into consideration.

To begin with, the question as to what the scope of ethics related to AAL is requires due consideration. To the knowledge of the author, there is no comprehensive or exhaustive compilation or analysis of ethical matters in this regard. Some publications try to give an overview of this important field and may serve as a starting point [Borges et al., 2008, Kubitschke et al., 2009].

Ethical issues affect all stakeholders concerned with AAL in its broadest sense: Researchers and developers, designers, healthcare and social workers as well as their umbrella organisations, sociologists, and policy makers at governance level should understand that the overall aim of AAL is serving for the good of older people, their family members, and the common good in general [Kubitschke et al., 2009]. In order to protect and safeguard the privacy and dignity of the elderly users, however, sophisticated guidelines are required. Table 2.2 summarises numerous ethical aspects that were identified and are being discussed among the players in the field of AAL at the time of writing this thesis (2009/2010).

During the design and development of assistive technologies, however, concerns about ethical questions may not prevail in order to prevent that new approaches are stifled in the first place. Seeing technological development only two sided (safety for the user vs. civil liberties) is no viable way for tackling the challenges connected thereto. Examples for AT projects that turned out to be beneficial both for the users ("patients") and the caregivers are described in technical literature, e.g., a project involving boundary alarms to monitor and track people who wander. In this particular case, the system fulfilled the expectations and reduced stress both in patients and caregivers [Hughes & Louw, 2002].

Apart from the impact of AT on the user, their self-determination, and their privacy, it is of crucial importance to distinguish *therapy or assistance* and *enhancement*. *Therapy or assistance* means the attempt to restore capabilities that dwindled away and to compensate for mental or physical handicaps up to a natural level whereas *enhancement* describes efforts to improve human capacity beyond that of a healthy person. The distinction, however, is not trivial [Bostrom & Roache, 2007]. In case of many a medical treatment, it is not immediately clear what it is to be classified as (e.g., vaccinations). Moreover, human capabilities are not constant over lifetime. Thus, the question arises whether it is rather therapy/assistance or enhancement to render an 80-year-old mentally and physically capable to the level of a 20-year-old. As a rule, it needs to be determined in each individual case which medical condition or impediment warrants what kind of (AT) assistance [Manzeschke, 2009]. This brief paragraph, however, can only be meant to encourage further reading regarding the broad field of ethical questions connected to human enhancement (e.g., [Allhoff et al., 2009, Bostrom & Roache, 2007, Bostrom & Sandberg, 2009]).

Table 2.2: Ethical aspects relevant for AAL research and development (based on [Borges et al., 2008, Mahoney et al., 2007]

Informed consent	Informed consent is of paramount importance for any trial or usage of AAL technology. Informed consent means that the person giving their consent is capable of fully understanding the implications the usage of technology may have on their life. Moreover, the person must give their consent freely and without applying any kind of force or pressure.
Transparency	The user is to be treated fairly, i.e., he has to be informed fully and honestly about potential benefits, drawbacks, or privacy invasions connected to the usage of AAL technology. Truthfulness with regard to the implications of AAL usage is of pivotal importance.
Justice	Distributional fairness is to be guaranteed. AAL technology has to be provided to all social classes, regardless of wealth or education. Neither may anyone be deprived of AAL nor may anyone be pressured to accept it.
	The formation of a two-tier society in which either of the following two conceivable scenarios might become a reality must be avoided: Pressure for the poor to use AT because it is cheap but decent, personal face-to-face health care is not affordable *vs.* exclusive use of AT by the wealthy since they can afford the best medical care available.
Respect	Respect for prospective users is crucial in a plenitude of ways:
	First, the individual who is going to use AAL technology must be respected no matter what their personal health status may be (disabled, cognitively impaired, suffering from dementia, etc.). This includes respect for personal rights, decisions, and preferences the individual may choose to exercise. As a rule, the integrity of the users may not be violated.
	Second, respect must also be paid to the family and caregivers of the affected individual. Neither of them may be urged to act in an undesired way or have their wishes ignored.
	Third, AAL technology must be applied in such a way that it respects family relationships. The use of AAL may not jeopardise existing care arrangements.
Autonomy	The demand of autonomy complements the aspect of respect. Every user has the undeniable right not to accept parts of an AAL system or reject it as whole. In addition, the user must be allowed to disable the system temporarily without having to justify their decision.
Proportionality	Any AAL solution must be as unobtrusive and unrestrictive as possible. The effort for the user to understand and use the system is to be minimised. Only the kind and amount of data to reach a specific goal –to which the user must have given their consent– may be collected. The balance between the attempt of avoiding harm and ignoring or overruling user intentions needs to be maintained.
Beneficence	The baseline of any and all AAL concept must be that it serves for the good of the individual who uses it as well as their family members. Technological developments solely aiming at reducing costs for public health schemes or improving working conditions of medical staff without benefits for the user ("patient") are to be rejected. The exploitation of the elderly or people suffering from chronic illnesses is not acceptable.

It can be concluded that there is no silver bullet solving all potential ethical issues arising in the wake of assistive technology. Until reliable guidelines and standards will be established, every research project as well as marketing novel technologies to end users requires careful individual assessment to minimise undesirable impacts on the users while –at the same time– making sure that technological advancement is not hampered.

2.4.2.   Privacy Considerations

In addition to ethical concerns whose implications may not be too obvious at first sight, privacy is another matter that needs serious attention. This becomes very clear when thinking about possible data losses or misuse of information obtained by means of AT. Data collected by AT systems are among the most sensitive and personal kinds of data of an individual since they represent detailed information about the health and mental status.

In order to protect this data from both wilful and unintentional disclosure, loss, modification, or misuse, stringent security measures need to be implemented. Among the aspects that need to be taken into consideration are the ones listed below (based on [Borges et al., 2008]):

+ **Lawful processing and nondisclosure of data:**

   National law of the respective countries needs to be obeyed. No use of data outside the legal framework is allowed. As a result, national and international jurisdiction should feel encouraged to monitor technological advancements closely and react if necessary. Moreover, the data may not be disclosed to anyone except for the staff members of the organisations with whom the users agreed to share the data. This is particularly important to prevent third parties that may have illicit interests from processing the data.

+ **Limited purpose:**

   The collected data may only be used to reach the purposes and goals as stipulated in an agreement with the users. Use of data outside this agreement is not permitted.

+ **Relevance and necessity:**

   Only the amount and kind of data that is necessary to fulfil the goals that were agreed upon with the users may be collected. Excessive collection of data is to be avoided.

+ **Accuracy and security:**

   The accuracy and integrity of the data must be guaranteed at all times. This is to ensure that no intentional manipulations of the data can occur. They might lead to erroneous interpretation results and ultimately cause detrimental actions or conclusions for the user.

+ **Limited storage:**

   The collected data may not be stored longer than necessary to fulfil the monitoring task of the AT system. After that period, they have to be deleted securely.

+ **Obeying user's rights:**

   In the process of collecting, processing, and storing the data, the right to informational self-determination of the user, general human rights, civil liberties, and additional agreements between the AT system operator and an individual may at no point be infringed upon. This includes that the data be deleted upon the express wish of the user.

When paying attention to the above points, data collection and storage should not give rise to ethical or privacy concerns and should be in accordance with all applicable law. It should thus be in the interest of any AT system operator to strictly follow the above guidelines to avoid distrust and suspicion from the users and to comply with ethical fundamentals.

3.   The Process of AAL Research and Development

3.1.   Introduction

In this section, the research and development process that was followed in this thesis will be explained. As set forth in section 1.2, one of the main goals of this thesis was designing an AAL solution that can be successfully deployed to the end user. There is not a single key to achieving this objective since multiple players' interests must be satisfied. The three most important stakeholders are the prospective users, their family, and the investor ultimately funding the installation of AAL technology. The latter can either be the user himself or third parties interested in providing this technology to someone, e.g. family members, insurance companies, or housing societies.

It is, therefore, most important for any AAL technology to enjoy complete confidence of the actual users as well as all other parties involved. To gain this trust, all stakeholders must be certain that the use of the AAL technology is mutually beneficial. First, the users need to be convinced that AT is capable of helping them to live independently while, at the same time, not unduly invading their privacy. Second, family members want to rely upon the systems so that they have the peace of mind that their loved ones are well taken care of. Third, the investor who actually funds the installation of AT must be sure of the investment paying off within a reasonable period of time.

In order to live up to all stakeholders' expectations, AT needs to meet a lot of requirements. On the one hand, AT must be flexible (e.g., being suitable for a large variety of structural conditions), competitively priced (i.e., using off-the-shelf components available at reasonable cost), and reliable (i.e., technically mature) to satisfy the investor. For the users, other criteria are important: The AT system must be transparent (i.e., the operational principle must be clear to the users), be easy to use, unobtrusive, non-stigmatising, and –of course– reliable.

Development methodologies that were applied to the presented AAL solution on top of the above basic design principles will be elucidated in the following two sections. Particular attention was directed towards analysing users' needs and wants as well as involving them in the development and design process to make sure that the AAL solution indeed fulfils their expectations.

3.2.   Analysis of Prospective Users' Requirements towards AAL

3.2.1.   General Considerations

A diligent analysis of the users' needs and wants is crucial if a successful and universally accepted AAL solution is to be developed. This holds particularly true since the cohort of aged people is still today often perceived as a homogeneous group rather than a highly diverse population stratum (see section 2.1) and stereotypes are widespread [Georgieff, 2008]. In the nineteen nineties, it had been reported that the perception of aged people was characterised by strongly polarised thinking: Either seniors were regarded as old, frail, and ailing, or they were depicted by clichés such as vitality, prosperity, or experience of life [Bruns et al., 1999].

By now, a more nuanced perception has been established in some scientific communities but is not yet commonplace, especially not among the engineering sciences. Scientific evidence indicates that, for instance, the decline of physical or mental capabilities or the onset of frailty is not coupled with a specific age. These declines of personal abilities are rather dependent on the lifestyle a person led in earlier phases of their life. Another issue is the number of people living single at different ages. The percentage of single-person households rises up to 53% in the age group of the 80- to 85-year-olds. As a result, a large proportion of them face isolation and loneliness. Moreover, females enjoy a higher life expectancy than males. Thus, the gender ratio shifts towards female seniors [Georgieff, 2008]. Because of the generally very high life expectancy among today's seniors, the terms *young old*, *old old*, and *oldest old* were introduced to illustrate the vastly varying conditions and expectations of this age group. In conclusion, it can be stated that the cohort of the people aged 60+ is by no means homogeneous. Thus, depending on the target user group, AAL solutions may need to be personalised and highly adaptable to the specific demands.

The above considerations underpin the necessity to carefully assess the requirements of seniors towards AT. Needs and wants can and will change as people get older. Engineers designing AT, however, tend to follow technology-driven approaches and evaluate the usability of the systems. This approach is a technologistic one not considering how technology will be embedded in older users' everyday life. If AT fails to meet the users' real needs, any and all attempts to make users accept them will be futile and render the AT irrelevant [Sixsmith, 2009]. Therefore, it needs to be kept in mind at all times that the paramount objective of AT is enhancing independence and quality of life of the prospective users in a way they are content with.

Furthermore, it is important to note that age-related "problems" potentially perceived by researchers and developers of AT may not be as relevant to seniors as younger people think they are. This aspect is especially important regarding solutions for problems that emphasise age-related deficiencies. Seniors tend to refuse "help" or "solutions" that they perceive as being stigmatising, i.e., assistive devices that visitors can see and that show them the dependence of their users (such as patient lifts for bathing). Thus, many a simplistic approach identifying "obvious" problems, deducing "objective" needs based thereupon, and devising solutions thereto had been doomed to failure. In addition, seniors may develop unexpected or unconventional strategies to compensate for age-related deficiencies that they prefer over solutions offered by engineers or medical staff [Sixsmith, 2009]. Some examples for such coping strategies are cited in [Brennan & Cardinali, 2000], e.g., relying on memory and touch in order to compensate for visual impairments or simply using chairs when performing tasks that used to be done standing.

3.2.2.   Initial Requirements Analysis

Extensive research has been carried out in collaboration with the department of urban sociology (headed by Prof Dr ANNETTE SPELLERBERG) to identify the needs and wants of seniors. In [Spellerberg et al., 2009], an overview of multiple reasons contributing to the desire to use AT was compiled: First, health aspects, the prevention of illnesses, and care schemes based on telemedicine are named as incentives for using AT. Second, seniors' households are well equipped with appliances and ICT. Thus, seniors feel confident to use and master modern AT. Third, living independently at home rather than requiring institutionalised care is very important to a vast majority of 80% of all seniors. Hence, AT will be accepted provided that it is affordable. Fourth, today's seniors are mostly quite cost-conscious. Thus, many of them hope to be able to save costs for lighting and heating by using AT.

Based on the above findings, a preliminary feature set for the future AAL solution was defined. Since it is believed to be pivotal to stimulate prospective users of AAL technology to use AT as early as possible so that they get acquainted with it well before they need the health monitoring functions, a broad-scaled initial concept was drawn up. This initial concept comprises functionalities in the realms of comfort, safety, health, and communication/entertainment (see Table 3.1). By implementing many more functionalities than just health monitoring, both stigmatisation and lack of interest in the AT were to be avoided.

Table 3.1: Functionalities of the initial version of the devised AAL system

Comfort	Assistance in daily life	• Electrically driven roller blinds • Automatic lighting • Remote door opener • Alarm clock
Safety	Securing against internal and external threats	• Video entry phone • Visitors' history (saved images) • Switching off electric loads • Prevention of water damage
Health	Detection of emergencies and provision of help	• (In-)Activity monitoring • Alarm generation rules • Fall detection • Multi-level alarming scheme
Communication/ Entertainment	Provision of modern, ICT-based communication and entertainment	• Web radio • Internet bookmarks • Television

The feature set outlined in Table 3.1 was chosen because preliminary opinion polls among the future users indicated that they are capable of handling these functions. The comfort functions aim at easing daily life by using modern home automation technology for everyday tasks. Safety functions are meant to safeguard the user and their home but without relation to health. This includes protection from intruders or doorstep fraud as well as preventing damage to the flat caused by water, fire, etc. This can, e.g., be achieved by switching off dangerous electrical devices –such as electric irons or kettles– by pushing a button next to the front door. At this point, it needs to be noted that many of the above functionalities contribute to more than one of the four main assistance realms. The video entry phone in connection with the remote door opener, for instance, means comfort but also contributes to the safety of the user by storing a picture of the person seeking entrance.

With regard to health, activity monitoring was the main function to be achieved. Activity monitoring can be accomplished by merely using ambient sensors and does not require the identification of activities of daily living (ADLs). Thus, it does not infringe on the privacy of the users. It was assumed that both long-term changes in the daily routine in the range of weeks (possibly pointing at changes in the overall health level) and medium-term changes requiring soon intervention (in the order of minutes to few hours) can be detected (see Table 2.1). As explained above, it was never intended to be able to detect life-threatening emergencies of type A that necessitate immediate action. Based on activity pattern, rules were to be established governing the triggering of alarms and the alarm sequence, i.e., to whom and in which order (friends, relatives, emergency response service) an alarm is reported.

Last but not least, communication and entertainment were included as additional incentives for the user to adopt the technology as early as possible. Television and (web) radio are established forms of entertainment, whereas Internet bookmarks are novel kinds of media for most users.

3.3.   Development Process based on User Involvement and Feedback

As outlined in section 1.2, many a time the classic waterfall software development model (see Fig. 1.1) had been and still is being applied when designing software products. Already back in the nineteen eighties, however, the deficiencies of the waterfall model had been pointed out: "Software development is a complex, continuous, iterative, and repetitive process. […] [A linear] model […] does not reflect this complexity. In addition, the model does not reflect the parallel and iterative activities […]" [Wong, 1984].

Moreover, the waterfall model cannot accomplish user involvement in the way it had been implemented in this work. In Fig. 3.1, the amount of user integration is displayed for both the waterfall scheme (a) and the iterative model (b). In case (a), user feedback is only possible to obtain during the requirements analysis phase (i.e., even before software development has started) and during the testing phase (i.e., after software development will be finished). If any flaws or shortcomings are found in the software product after it has been released, it will be very difficult to fix them as this is not arranged for.

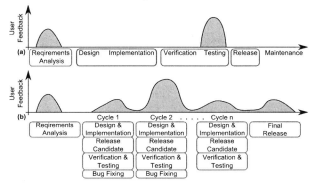

Fig. 3.1: Degree of user involvement and participation attained in case of linear waterfall software development (a) and iterative software development (b) (based on [May & Zimmer, 1996])

In contrast to that, multiple iteration loops are scheduled in case of the iterative development model (b). As in case of the waterfall model, a requirement analysis and a final test before the release are conducted. In addition, however, an unspecified number of iterative loops will be undergone. In this context, *unspecified number* means as many times as necessary to achieve the stipulated goal and the desired quality and functional range of the software. Each of the iterative loops comprises the steps *design & implementation*, *release candidate*, *verification and testing*, and *bug fixing*. That means that from step *n* to step *n+1*, only a limited amount of new functionality is added to the software. The resulting interim version is then thoroughly tested and, subsequently, released to the users. If any bugs or malfunctions should be found, they will be fixed in the next interim version. By following this development scheme, continuous user feedback can be ensured.

As a result, the software is constantly being altered, improved, and enhanced. At the same time, the functional range can also be truncated when verification and testing show that a function is superfluous or undesired. Only to mention an actual finding that will be discussed in more detail in a later chapter, several users wanted to have the automatic lighting in the vestibule of their flats disabled. The reason for this wish was that people disliked the lighting automatism but rather wanted to decide for themselves whether, when, and for how long the light should be on. This wish was in perfect accordance with the guidelines stipulated for this project that define autonomy of the user as one of the paramount design principles. By contrast, other features such as an electronic bulletin board ("forum") were demanded by the users. These two examples of user feedback are illustrative how the tenants grew from participating in this AAL pilot project: Even though they never had formal education in the fields of computer science or technology development, they can draw from their wealth of experience of life. Using this rich experience, they were able to contribute valuable suggestions for the further development of the AAL system.

In summary, this development scheme can be called a "human centred approach" (HCA). Essentially, HCA means integrating procedures for ensuring the usability of a software product into the software engineering process [Ferre & Medinilla, 2007]. In the AAL context, however, HCA means even more than that – not only is it desired to ensure usability but to guarantee the participation of every user throughout the entire design process. By going through several development cycles, the software is continuously being refined. Thus, user participation during all stages ultimately facilitates both technology acceptance and the sophistication of the resulting technology.

4.   Sensor Data Collection and Processing

4.1.   Why is Collecting Redundant Data Important?

The emergency detection algorithms presented in this work are based on information gathered by numerous, redundant, ambient sensors in the surrounding of a person. Since off-the-shelf sensors for building automation are being used, each individual sensor is not of industrial quality, i.e., not highly available. Failure rates, however, are still very low – within two years, only very few malfunctions became known among more than 600 modules used in a pilot project. Hence, it can be assumed that using multiple sensors is suitable for identifying average long-term activity or inactivity patterns of a person as well as monitoring the current activity or inactivity level.

The multiple data collection and processing steps involve recording, transmission, conversion, condensation, interpretation, and storing of the information. During this entire procedure, the data is continuously being refined to allow increasingly sophisticated conclusions about the tenant and their environment (see Fig. 4.1). This can be achieved by gaining further knowledge and eventually wisdom beyond the information contained in the initial raw sensor data. In the following, this process will be elucidated in detail.

4.2.   The Data Processing Algorithms: From Raw Data to Tertiary Information

4.2.1.   Preliminary Remarks

The internal design of the data processing procedure followed in this thesis is illustrated in Fig. 4.1. Four different types of data/information are distinguished: Sensor raw data, primary information, secondary information, and tertiary information.

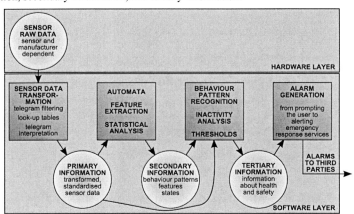

Fig. 4.1: Collection, processing, and interpretation of the sensor data

This chapter addresses the algorithms devised for obtaining tertiary information about health and safety using advanced inactivity analyses. Before this attempt can be made, however, primary and secondary information need to be extracted from the initial sensor raw data. The required steps are described in the sections 4.2.2–4.2.4.

4.2.2.   Sensor Raw Data

Sensor raw data is the original data as provided by the ambient sensors without any processing having taken place. In the AAL context, several types of sensors can be used to monitor the users. There are, however, two fundamentally different classes of sensors: First, there are genuine sensors of various kinds whose only function is capturing data and, second, quasi-sensors whose primary use is not gathering information but being used for controlling the home environment. Examples for quasi-sensors are, for instance, wall switches that are connected to a data bus so that information about when they are actuated can be retrieved. Thus, normal interaction of the user with his environment also generates information about his location and activity level.

Moreover, it should be noted that it will be important in the following to differentiate between *singular events* and *continuous activity*. The former are triggered by momentary events (all sensors but the motion detectors) whereas the latter denotes phases of activity while at least one motion detector is in its active state (see section 5.2 for details).

Details about the used home automation components will be discussed in chapter 8, but Fig. 4.2 already gives an overview over the amount of data collected in each flat in the Kaiserslautern project. Several hundred or even thousand activity events are observed every day in a flat. In the below sample diagram, approximately 2500 events are displayed that were captured in a single day. Without going too deeply into which sensors are installed in the flats, this diagram illustrates the frequency and variety of activity events very well. The large gap at night without any activity apart from the alive signal of the AAL system (i.e., the equally spaced line at the bottom of the diagram) was caused by the tenant having been asleep from 0:30 a.m. to 6:00 a.m.

Since the algorithms and methodologies developed within the scope of this thesis are to be as generic and universally usable as possible, dependencies on specific sensor models and manufacturers or proprietary protocols are to be avoided. As a result, sensor raw data is highly unstructured. They cannot be processed by the data interpretation algorithms working on primary, secondary, or tertiary data without first being transformed and standardised.

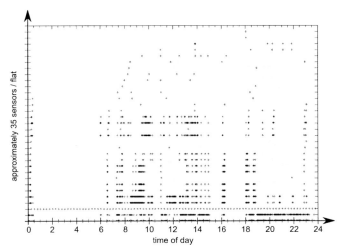

Fig. 4.2: Raw data collected over 24 hours in flat A

Within this work, several sensors provided by various manufacturers had been evaluated. These tests revealed that the protocols, data structure, and the encoding can be entirely different. Two different kinds of data telegrams sent by two motion detectors from different companies are shown in Fig. 4.3.

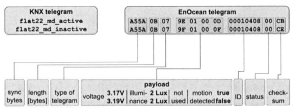

Fig. 4.3: Two examples for manufacturer-dependant sensor raw data

The above figure illustrates the entirely different nature of the telegrams from different sensor manufacturers. In case of the sensors compliant to the KNX standard, the telegrams are provided as plain text, i.e., they are static and contain the payload in human-readable form ("flat22_md_active" = "motion detector in flat 22 becomes active") whereas the EnOcean telegram contains much more information than only whether or not motion had been detected. EnOcean modules are solar-powered and transmit their battery level and the illuminance as well[1]. Thus, the contents of the telegrams are highly dynamic and need to be decoded before they can be interpreted in subsequent steps.

---

[1] For all reasoning described hereafter, only the information about activity will be considered. All other information encoded in the EnOcean telegram is irrelevant for this work.

4.2.3.   Primary Information

Depending on the sensor model used, the frequency of the telegrams can vary considerably (Fig. 4.4). In the above example, the KNX sensors only signal rising and falling edges, i.e., send telegrams when a state changes (e.g., motion starts or ceases). In contrast to that, the EnOcean sensors send routine telegrams about their current state approximately every 1000 seconds. In addition to those routine telegrams, edge transitions (i.e., state changes) are transmitted as well. Thus, the received telegrams need to be filtered to make sure that routine telegrams are interpreted differently from actual edge transitions. It needs to be noted that the delays between a state change and the telegram notifying of this state change, however, vary between 0 and 100 seconds. If only activity and inactivity patterns are to be determined, these delays are rather unproblematic but if location tracking of the tenant is to be performed, this behaviour of the sensors can complicate the evaluation considerably (location tracking and details on this phenomenon will be described in the section *advanced alarming criteria* below).

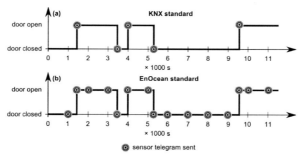

Fig. 4.4: Edge transition signalling only vs. edge and routine signalling

Moreover, the fundamentally different data structure depicted in Fig. 4.3 renders it necessary to decode and transform the telegrams sent by the sensors so that all subsequent data processing algorithms can handle the information, no matter with which type of sensor it had been collected. All algorithms operating on primary, secondary, or tertiary data that were implemented in this work expect generic, unambiguous, well-defined strings (e.g., `front_door_open`) that are *not* specific to a particular flat or AAL installation. Once the transformation of the telegrams has been completed, the standardised telegrams are passed on to subsequent processing steps.

In case of KNX telegrams, the necessary transformation is straightforward. The flat-specific telegrams are simply looked up in a dictionary ("look-up table") linking the specific telegram with a generic one, e.g. `flat22_md_bedrm_active` would be converted into the generic telegram `motion_detector_bedroom_becomes_active`.

Decoding EnOcean telegrams is slightly more complex. Two corresponding telegrams (e.g., *motion detector active* and *motion detector inactive*, respectively) do contain more information than only the actual state of the sensor (Fig. 4.3). In case of the motion detectors, for instance, the battery and illuminance level are also encoded in the telegram. Thus, a simple look-up table is no viable way to convert them into the desired generic telegrams. In this case, the received telegram needs to be parsed (i.e., interpreted) to extract the required information from it. Depending on the parsing result, the respective generic string will be generated.

4.2.4. Secondary Information

Secondary information is the first type of information that consists of aggregate, condensed information based on data from previous processing steps. In order to obtain secondary information, primary data has to undergo multiple steps of further refinement, the first step being running finite state machines (FSMs, automata) on the data. In order to make the primary data available to the FSMs, an *observer design pattern* has been implemented in which the standardised telegrams raise an event that can be subscribed to by the FSMs. The FSMs, their structure, and which events each FSM has to subscribe to are defined in a configuration file, rendering the definition or modification of FSMs easy and flexible.

The FSMs are used to model the states of the individual sensors, e.g., the information about doors (open/closed), windows (open/closed), lights (on/off), and roller blinds (up/down). More importantly, FSMs are suitable for deducing knowledge that is not immediately obvious from raw or primary data. Two FSMs for obtaining specific knowledge will be presented below in the appropriate sections. Amongst the most fundamental and important FSMs are the ones for determining whether or not there is any activity inside the flat ("*activity FSM*") and whether or not the flat is occupied by a human ("*presence FSM*"). These two FSMs are interdependent, i.e., the *presence FSM* is dependent on the *activity FSM* in order to work properly. The concept of activity and inactivity will be described in detail in chapter 5. In addition to the *activity FSM*, the *presence FSM* is also dependent on the output values of the FSMs representing the individual (quasi-)sensors – e.g., wall switches or other objects the user can interact with. These FSMs generate the output value singularEvent when a transition is performed and thus signal activity as well. The FSM modelling the door generates output signalling that the door was opened or closed. Implementation details of the FSMs are illustrated below on the basis of the *presence FSM* which is an excellent example for the FSM design (see Fig. 4.5).

All FSMs used in this work are based on Mealy automata. Another automaton type that could represent the automaton structures needed in this work is the Moore automata. The main difference between Mealy and Moore automata is that the output of the automaton is determined by the actual transition (Mealy) rather than the destination state of a transition (Moore). However, since Moore automata can easily be transformed into Mealy automata [Datta, 2001] and Mealy automata offer significant advantages regarding the actual C# software implementation of the automata, Mealy automata were preferred in this work. The Mealy automata used here are a 6-tuple of the following nature (4.1)

$$A_{\text{MEALY}} = (\boldsymbol{Z}, \boldsymbol{V}, \boldsymbol{W}, G, H, z_0) \tag{4.1}$$

where $\boldsymbol{Z}$ is the set of states, $\boldsymbol{V}$ is the input alphabet, $\boldsymbol{W}$ is the output alphabet, G is the transition function, H is the output function, and z_0 is the initial state of the automaton [Lunze, 2007].

The transition function G_{MEALY} and the output function H_{MEALY} are defined as follows (4.2).

$$\begin{aligned} G_{\text{MEALY}} &: \boldsymbol{Z} \times \boldsymbol{V} \to \boldsymbol{Z} \\ H_{\text{MEALY}} &: \boldsymbol{Z} \times \boldsymbol{V} \to \boldsymbol{W} \end{aligned} \tag{4.2}$$

Equations (4.2) illustrate that in a Mealy automaton, a transition can be performed only when an input v is received that enables a valid, possible transition $G_{\text{MEALY}}(z, v)$ out of the currently active state z. The output function $H_{\text{MEALY}}(z, v)$ defines the corresponding output value w for each transition $G_{\text{MEALY}}(z, v)$ and is also dependent on the currently active state z and the received input v.

For the purposes that are being pursued in this work, however, standard Mealy automata are not sufficient. In many cases, it is mandatory that time constraints are implemented which control whether or not a transition may be performed. Hence, the automata definition in (4.1) needs to be modified to allow for the option to set individual time constraints and clocks (similar to stop watches) for each transition. Equation (4.3) shows the expanded general automata definition used in this work.

$$A_{\text{AAL}} = (\boldsymbol{Z}, \boldsymbol{V}, \boldsymbol{W}, G, H, z_0, \boldsymbol{C}, \boldsymbol{T}, \boldsymbol{c_0}, \boldsymbol{R}, M), \ \boldsymbol{R} \subseteq \boldsymbol{V}, \ \|\boldsymbol{C}\| = \|\boldsymbol{T}\| \tag{4.3}$$

where $\boldsymbol{C} \in \mathbb{R}_0^+$ is the set of clocks, $\boldsymbol{T} \in (\mathbb{R}_0^+, \mathbb{R}_0^+)$ is the set of time constraints (i.e., time spans), $\boldsymbol{c_0}$ is the set of initial clock conditions, $\boldsymbol{R}$ is the set of input values (in the AAL context called *telegrams*) that reset the clocks to zero, and M is the mapping function defining which telegram resets which clock. The set $\boldsymbol{R}$ of telegrams resetting the clocks is always a subset of the input alphabet $\boldsymbol{V}$. The number of time constraints is always equal to the number of clocks. A special time constraint of type 0, the "empty" timer t_0, meaning that transitions controlled by it are not subject to any time restrictions, is always element of $\boldsymbol{T}$. One time constraint t may control more than one transition.

Consequently, the transition function G_{AAL}, the output function H_{AAL}, and the mapping function M_{AAL} are defined as follows (4.4).

$$G_{AAL} : \boldsymbol{Z} \times \boldsymbol{V} \times \boldsymbol{C} \times \boldsymbol{T} \times \boldsymbol{R} \to \boldsymbol{Z}$$

$$H_{AAL} : \boldsymbol{Z} \times \boldsymbol{V} \times \boldsymbol{C} \times \boldsymbol{T} \times \boldsymbol{R} \to \boldsymbol{W} \qquad (4.4)$$

$$M_{AAL} : \boldsymbol{R} \to \boldsymbol{C}$$

Thus, certain timer-controlled transitions $G_{AAL}(z, v, c, t, r)$ may only be performed if a specific telegram v_{reset} had been received within a certain time frame $\tau \in t$ with $t \in \boldsymbol{T}$ before the telegram v triggering $G_{AAL}(z, v, c, t, r)$ so that the clock c has been reset and now allows the transition to be executed.

In Fig. 4.5, the *presence FSM* is displayed. This FSM determines whether or not there is human presence inside the flat. It consists of four states and eleven transitions. Two time constraints are implemented, one of them being the empty timer. The empty timer is symbolised by a crossed out hourglass (⌛). In addition, the FSM accepts six input values and can produce three output values (`occupied`, `unoccupied`, `presenceUnclear`). Considering the various components and capabilities of the FSMs, it needs to be pointed out that the two main purposes of the FSMs are on the one hand representing real-world conditions of the modelled system by means of its respective active state and on the other hand generating output values. The output values constitute part of the newly obtained secondary information which cannot be determined using sensors.

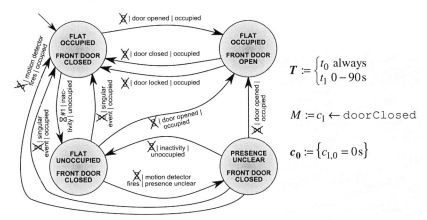

Fig. 4.5: Finite state machine (automaton) for determining human presence inside the flat[2]

Three examples shall exemplify the operational principle of the *presence FSM*.

+ Transition from [flat unoccupied | front door closed] to [flat occupied | front door open]: This transition is not subject to any time constraint, i.e., it may be performed any time the telegram `doorOpened` is received. Upon reception of this telegram, the transition will be triggered and [flat occupied | front door open] will

---

[2] The labels of the transitions consist of three parts: *timer | input value | output value.*

become the new active state. "Front door open" is a direct result of the action "door opened", whereas the knowledge "flat occupied" must be reasoned – if the flat is empty and the door is opened (not assuming that a burglary is taking place), it makes sense to reason that the flat will be occupied from that moment. Since the knowledge to be generated by this FSM is whether the flat is occupied or not, this knowledge is encoded in the output value of the transition, allowing the entire AAL system to adapt its actions and behaviour accordingly (i.e., most importantly not raising alarms if no one is at home).

+   Transition from [flat occupied | front door closed] to [flat unoccupied | front door closed]: At first sight, this transition does not seem to be justified. Upon closer inspection, however, this transition is indispensable. The motion detectors switch off with a delay of 12-90 s after the last activity has been observed. Thus, it is possible to leave the flat and close the door while some of the FSMs modelling the motion detectors are still in the state *active*. Hence, a transition subject to a timer of 0-90 s (reset by the telegram doorClosed) had been introduced, meaning that in a time interval of up to 90 s after closing the front door, the active state can be changed to [flat unoccupied | front door closed] if any and all activity ceases inside the flat, including activity possibly signalled by other sensors than motion detectors (e.g., pressing switches or interacting with doors or windows). Applying a timer to this transition is crucial to prevent the transition from being triggered accidentally if there is a phase without any activity but the door not having been closed shortly before.

+   Transition from [flat unoccupied | front door closed] to [presence unclear | front door closed]: The state [presence unclear | front door closed] has been implemented to improve the robustness of the FSM. Since situations are conceivable in which the timer-controlled transition described above may lead to faulty reasoning and thus to erroneously reached states, correction procedures must be implemented to allow for the rectification of faulty reasoning. In the course of this work, however, it turned out that motion detectors may erroneously detect motion if the low afternoon sun shines on them. Hence, simply regarding activity sensed by a motion detector is no sufficient evidence for human presence. When in the state [flat unoccupied | front door closed], the FSM will switch to [presence unclear | front door closed] upon the reception of the telegram motionDetectorActive. If a second motion detector fires as well or if a singular event occurs, the state [flat occupied | front door closed] will be reached, whereas the FSM will switch back to the previous state [flat unoccupied | front door closed] as soon as the first motion detector signals inactivity again. On a side note, it needs to be mentioned that the *presence FSM* will immediately switch from [flat unoccupied | front door closed] to [flat occupied | front door closed] upon reception of the telegram singularEvent – telegrams other than from motion detectors are an unmistakable indication of human presence.

In summary, amongst others the FSMs generate new knowledge about features such as presence, overall activity, etc. that cannot be observed directly by physical sensors. These features, however, are indispensable for further data processing. Both the original primary information and the automata output are then statistically analysed to determine activity and presence/absence patterns that constitute the basis for reasoning tertiary information on health and safety of the users.

4.2.5. Tertiary Information

Obtaining tertiary information is the main focus of this work and revolves round the techniques of behaviour pattern recognition, inactivity analysis, and threshold application. In these steps, high-level information about health and safety of the tenant is deduced. These techniques are elucidated in full detail in the following three chapters.

5.   Activity vs. Inactivity Approach

5.1.   Introduction

The principal aim of this work is developing an AAL system that is capable of detecting medical emergencies arising among its users. As illustrated in Table 2.1, three different classes of emergencies need to be distinguished. Since the AAL approach followed here is genuinely ambient, i.e., not involving any sensors worn on the body, class A emergencies requiring immediate attention cannot be addressed. Instead, the AAL solution being developed within the scope of this work is ideally suited for detecting incidences of class B. Since some delay is inherently attached to this approach, class B emergencies can, however, only be noticed with a certain delay. Examples for class B emergencies are falls, faints, and the like. Emergencies of class B are typically also targeted at by conventional PERS's. To a degree, behavioural changes associated to class C (i.e., the deterioration of the user's state of health slowly manifesting itself) may be detected as well.

In the following section, the two different basic types of human actions encountered in AAL settings will be delineated. In addition, the notions of activity and inactivity will be introduced. They are of pivotal importance for monitoring the users and detecting class B incidents. It turned out, however, that inactivity monitoring is more suitable than activity monitoring for identifying class B emergencies; the advantages of evaluating inactivity rather than activity will thus be discussed as well. This is again in accordance with the concept of iterative development: At first, activity had been deemed to be the key to health monitoring but then inactivity proved to be the more viable approach. Emergencies of class C, however, cannot be detected using basic inactivity monitoring but require advanced interpretation methods which will be addressed in section 6.4.

5.2.   Singular (Transient) Events vs. Continuous (Non-Transient) Activity

As outlined above, most sensors in AAL environments detect *singular, transient events*. These events are, however, fundamentally different from *continuous activity* observed by motion detectors that represents a time span rather than a discrete event (Fig. 5.1).

The below illustration shows a sample string of telegrams that might be received when a person returns to their home, opens the front door and enters the flat (telegrams 1-4), resides in the sitting room[3] (telegram 5-8), and finally leaves the flat again (telegrams 9-11). Telegrams 12 and 13 are not actively caused by the user but are generated in the wake of the tenant leaving. The diagram illustrates the two different types of activity: Singular events, i.e., events that only represent a momentary action, are marked by black upward arrows. As a rule, all interactions of the user with their environment raise singular events – pressing buttons, opening doors, etc. rarely have a duration.

---

[3] In the flats considered in this work, the kitchen and sitting room are combined, similar to an open-plan apartment. For reasons of simplicity, this space will be referred to as *sitting room* in the following.

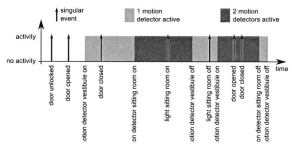

Fig. 5.1: Example of sensor signals and resulting types of activities (singular activity vs. time spans)

In contrast to the singular events, activity observed by motion detectors (grey areas in Fig. 5.1) is not transient but is continuous and lasts as long as a person is moving plus the switch-off delay, i.e., an arbitrary period of time.

Definition 5.1:	
singular event	event raised upon observable momentary interactions of the user with their environment
continuous activity	extended periods of continuous activity as observed by motion detectors

The inherently different nature of those two types of activity renders it very difficult to merge them. Fig. 5.1 shows that while at least one motion detector is active (grey areas), all information about singular events becomes redundant. On top of that, singular events should rarely be received outside continuous activity periods since motion detectors should detect human presence even without relying on singular events being raised. However, singular events can become very important on two conditions: First, they contribute to the redundancy of the sensor array and can still provide some information about the user if one of the motion detectors fails or the tenant is out of range of the sensor so that the sensor cannot "see" him. Second, correlating singular events and continuous human presence allows the development of advanced data interpretation algorithms that can distinguish between "normal" and "suspicious" situations. In normal situations, it is expected that a certain minimum number of singular events occurs per unit of time of continuous activity (i.e., motion detectors being active). There is reason to assume a strong correlation between singular events and continuous activity since it is hardly conceivable that a person does not interact with their environment for prolonged periods of time while at home. Suspicious situations are situations in which only one type of activity is observed. One the one hand, this can mean that only continuous activity is observed, hinting for instance at a fall of the tenant who keeps crawling on the floor. On the other hand, inconsistent activity patterns (e.g., continuous activity in the sitting room and singular events from a window in the bedroom) may indicate attempted burglary.

5.3.   Composite Singular/Continuous and Condensed Activity Patterns

As outlined in the previous section, merging the two different types of activity is not straightforward. In the above approach, each day is divided into 24 h × 3600 s/h = 86 400 s and a timeslot is assigned to each thereof. Using this scheme, activity patterns accurate to the second can be created. A sample of such an activity pattern recorded under real living conditions is shown in Fig. 5.2.

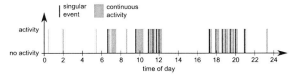

Fig. 5.2: Real-world sensor signals and the resulting singular and continuous activities

The main issue connected to merging singular events and continuous activity is that the "cardinality" of the observed activity has to be constant – either there is activity or there is none. Thus, overlapping activity signals from multiple motion detectors and singular events must be condensed into a single parameter, i.e., general activity. This can be achieved by constructing the envelope of all activity signals, no matter where they originated from (Fig. 5.3).

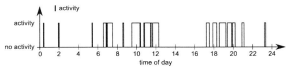

Fig. 5.3: Envelope (condensed pattern) of the activity graph shown in Fig. 5.2

One may be inclined to think that a possible approach for tackling the recognition of emergencies could be creating activity patterns –like the one depicted in Fig. 5.3– of each day for a particular flat and trying to extract features characterising the daily routine of a user from these patterns. Visualising and interpreting the recorded activity data as shown in Fig. 5.3, however, illustrates the two main drawbacks of this rather simple activity evaluation method: First of all, interpreting activity data in the above way hardly allows the detection of emergencies because –as a preliminary assumption– no emergency should have occurred as long as activity can be observed. I.e., in case of a genuine emergency, e.g., a faint or unconsciousness, all activity is expected to cease entirely[4]. Thus, comparing and matching activity patterns as shown above in order to detect emergencies is not very promising since focussing on activity is counterproductive if inexplicable or unanticipated *inactivity* periods indicate potential emergencies. This consideration leads to the approach of instead using inactivity as a criterion for detecting emergencies: If extended periods of inactivity cause any suspicion –be it because they are extremely long or at unexpected times– alarms can be triggered. In the

---

[4] Situations are conceivable, however, in which not all activity may stop. They will be addressed in section 6.4.

above figure, several examples of such inactivity periods are visible, notably before 0630[5], between 1200 and 1700, and between 2000 and 2400. Yet, the knowledge about these inactivity periods cannot be utilised directly for identifying dangerous situations and raising alarms because of the second drawback: The data representation in Fig. 5.3 does not include any information about the user being at home or the flat being empty. Without that additional knowledge, any attempts of reliable emergency detection will be futile because the AAL system cannot decide whether an actual emergency arose after a certain period of inactivity had elapsed or whether the user left their flat.

Thus, a sophisticated model for representing the users' inactivity is needed, taking into account whether or not the tenant is at home. Each model meeting this requirement will have to implement –amongst others– the presence FSM introduced in Fig. 4.5 which is capable of detecting human presence inside the flat. A suitable data interpretation methodology enabling such advanced inactivity analysis and recognition will be introduced in section 5.4.

There are, however, cases in which activity interpretation allows deducing specific knowledge not easily educible by inactivity analysis. Within this work, activity interpretation will only be performed as a means to improve and complement inactivity analysis, thus minimising the impact on the user's privacy. Such activity interpretation will include correlating singular events and continuous activity in order to corroborate any conclusions yielded by inactivity analysis (see 6.4). It needs to be noted that ADLs, one possible application of user activity data often encountered in the technical literature, will not be identified within this work. They allow profiling the daily routine of a person in detail and constitute a considerable privacy intrusion as not only activity and inactivity periods are monitored but specific actions, e.g., cooking, bathroom utilisation, etc. It is expected that long-term trends in a user's activity pattern –often tackled by monitoring ADLs– will also be reflected in the inactivity patterns so that gradual changes of the user's behaviour can also be recognised following the inactivity approach outlined above.

5.4.   Inactivity Patterns

In contrast to the *activity* patterns introduced in the previous section, *inactivity* patterns focus on the time *between* activity periods. This approach has several advantages over activity patterns. First, merging multiple sources of activity data is no longer necessary – inactivity is the case if *no sensor* whatsoever signals activity. Second, inactivity data interpretation as shown below (Fig. 5.4) provides additional knowledge – other than in case of the activity envelope in Fig. 5.3 that only shows whether or not there is activity at any given time, the inactivity representation also contains information on how long a person had been inactive at a given time.

In the schematic sample diagram in Fig. 5.4, four inactivity periods (chequered spikes) occurred throughout the day: Before 0800, 1000–1400, 1600–1900, and after 2200. Between

---

[5] In order to avoid any ambiguity when referring to the time of the day and to shorten the time notation, all times are given as "railway time" rather than using the 12-hour clock notation, i.e., 5:35 p.m. will be written as 1735.

each of them, activity periods were observed: 0800–1000 and so forth, i.e., each inactivity period commenced after activity ceased entirely and ended upon the reception of an arbitrary activity telegram. It needs to be noted that *activity* is *not* directly visible in the diagrams because activity means *no inactivity*.

When an inactivity period begins, a spike starts rising linearly from the abscissa. The height of a spike indicates the duration of the inactivity period, e.g., the duration of the inactivity period between 1000 and 1400 is four hours which is reflected by the height of the corresponding spike. Upon reception of an activity telegram, the inactivity period is terminated and the spike plummets back to zero.

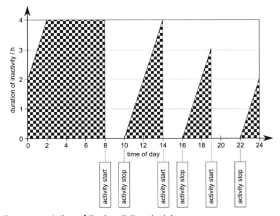

Fig. 5.4: Schematic representation of the inactivity principle

A formal expression for the duration of inactivity *DI* is given in Eq. (5.1)

$$DI(t,d) := \text{elapsed time after last observed activity}$$
$$t \in (0\,\text{h},\ 24\,\text{h}),\ \ d \in \mathbb{N} \tag{5.1}$$

where *DI* is a non-negative function, *t* is the time of the day, and *d* is the day considered.

For the third advantage of inactivity patterns over activity patterns to materialise, the inactivity monitoring procedure as illustrated in Fig. 5.4 needs to be expanded to include the *presence FSM*. Thus, presence and absence of the tenants can be accommodated in the inactivity approach which cannot be accomplished by the activity approach. The above activity graphs (Fig. 5.2 and Fig. 5.3) do not inherently include knowledge about the user being at home or not and may thus lead to faulty reasoning. In contrast thereto, statistical analyses of inactivity data as in Fig. 5.5 will not be distorted because the impact of absence can be quantified and can thus be compensated for.

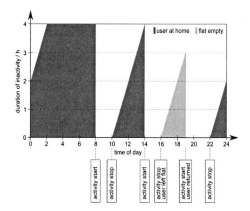

Fig. 5.5: Schematic inactivity diagram distinguishing between absence and presence

In the above figure, spikes coloured dark grey indicate inactivity while the tenant is at home whereas light grey spikes denote periods of inactivity due to the flat being empty. These two cases can be distinguished using the output generated by the presence FSM which in turn combines activity and door sensor telegrams to obtain this knowledge. In the above example, the tenant left their flat at 1600 and returned at 1900. Using the knowledge about the tenant being at home or away, Equation (5.1) can be modified to only represent inactivity while the tenant is present (see Eq. (5.2)). Periods during which the tenant is not at home do not need to be considered for detecting emergencies or building long-time inactivity patterns.

$$DI_\mathrm{p}(t,d) := \begin{cases} DI(t,d) & \text{if tenant present} \\ 0 & \text{elsewhere} \end{cases} \tag{5.2}$$

In the example in Fig. 5.5, DI_p is represented by the dark grey spikes. Only these indicate inactivity of the user that was not caused by absence. Applying the above considerations to the real-world data shown in Fig. 5.2 yields the inactivity diagram in Fig. 5.6. This diagram illustrates that the tenant was not at home during the long period of inactivity between 1200 and 1700. All other phases of inactivity can be put down to the tenant being rather passive, i.e., only moving little and not interacting with any buttons or sensor-equipped objects in their surroundings. Possible reasons for this include taking a nap or sitting on the sofa and reading a newspaper – if only little physical action is exercised, the presence sensors may not detect it.

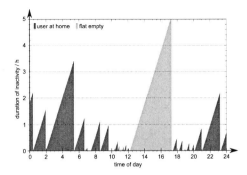

Fig. 5.6: Inactivity graph of real-world data shown in Fig. 5.2

A detailed discussion of inactivity patterns, further processing and interpretation steps, and the conclusions that can be drawn from them follow in the next section.

6.   Alarm Generation based upon Inactivity Analysis

6.1.   Introduction

Based on inactivity graphs that were introduced in the previous chapter, extensive analyses of the personal long-term inactivity patterns of a user can be performed. A sample inactivity pattern of a single day, however, will not suffice as reference, no matter what kind of emergency is to be identified. Thus, statistically more significant inactivity reference data sets need to be generated that well reflect the typical daily routine of a person. Long-term graphs created by combining inactivity patterns of multiple days, thus representing the average long-term behaviour of a user, are a suitable representation of the daily routine and are introduced in section 6.2. Subsequently, various thresholds will be deduced from the multi-day graphs that can serve as alarm criteria for emergencies of type B, i.e., alarms will be triggered as soon as the inactivity duration in a flat at any given time exceeds the alarm threshold at the respective time. Alarm thresholds can either be linear or non-linear as well as active 24/7 or only at certain times of the day. In section 6.4, advanced alarming criteria will be presented that may help to improve the detection rate and selectivity of the emergency recognition algorithms, reduce the number of false alarms, or enable the identification of slowly developing health issues of type C.

6.2.   Multi-Day Inactivity Patterns

6.2.1.   Preliminary Considerations

In order to generate long-term inactivity patterns from which reliable conclusions about the user's typical behaviour can be drawn, data from several days or even weeks must be consolidated into a single data set. This way, the significance and generality of the resulting inactivity data set will be increased. This step is imperative since only this kind of condensed information allows matching of the current inactivity pattern with the learned one. It needs to be noted that during this learning process no monitoring can be performed yet.

Several methods for condensing sets of single-day profiles are conceivable. In this section, the four types of multi-day patterns used in this work will be presented in detail. The first and most straightforward method is overlaying multiple single-day patterns and removing inactivity periods during which the user was not at home ("light grey spikes"), thus eventually representing maximum inactivity. Similarly, single-day patterns can be overlaid by calculating the mean inactivity rather than the maximum inactivity. Moreover, long-term inactivity patterns with statistical outliers being removed will also be introduced. Finally, a smoothing algorithm will be applied to the outlier-free long-term inactivity patterns.

All multi-day patterns that will be investigated in the following comprise data from 28 individual, consecutive days because it is assumed that 28 days constitute a reasonable learning period that is sufficient for observing the typical behaviour a tenant exhibits. Longer teach-in phases are deemed inappropriate because they would render the AAL system non-operational, i.e., not being able to monitor the tenant or to trigger alarms, during the learning phase. It must be noted in addition that all data evaluations and assessments in the following will be conducted on a per-month basis, i.e., if data of several months are evaluated, all months will be treated as entirely independent entities.

6.2.2.   Maximum-based Multi-Day Patterns

Fig. 6.1 demonstrates the calculation of a maximum-based multi-day inactivity pattern based on 28 days. Periods of time during which the user was not at home do not contribute to establishing a criterion for identifying dangerous situations and are thus omitted in the multi-day patterns: Only dark grey spikes represent user behaviour that was observed inside the user's flat. Among the main advantages of maximum-based multi-day patterns are that they are easy to compute and that they are readily comprehended by both professionals (e.g., developers or operators of the system) and end users. One of the drawbacks is that maximum-based patterns are unduly susceptible to exceptional one-time behaviour that does not represent the typical daily routine of a user ("outliers"). Such outliers may distort the long-term pattern to a degree that can render it unsuitable for being used as a reference pattern. In the following, maximum-based multi-day patterns will be discussed in detail.

The data shown below was collected in an individual's flat in a summer month. Two features can be demonstrated with the below diagrams. First, diagram (o)–(u) show days on which the user was not at home. On all of these seven days, however, short interruptions of the inactivity (i.e., short periods of activity) occurred. They can be traced back to two causes when looking into the raw data logs which contain detailed information on the type and number of the telegrams received throughout a day: On the one hand, the tenant has apparently entrusted a friend or neighbour with looking after their flat or plants. The raw data clearly indicates that someone entered the flat for short periods of time every day. On the other hand, some of the inactivity interruptions were apparently caused by malfunctions of the sensors. Dependability tests of the used sensors and actuators were conducted at the Institute of Automatic Control at the University of Kaiserslautern in order to verify that the equipment indeed conforms to the specifications as published by the manufacturers. These tests revealed that the KNX presence detectors used in the Kaiserslautern project (see section 8.3.2.2) also respond to sudden changes in intensity of illumination rather than motion only. Detailed analysis of the raw data corresponding to the below diagrams substantiates the assumption that presence detectors were erroneously triggered by low sun (see also section 8.5).

In addition to the above finding, the second feature that can be extracted from the below series of diagrams is a specific regularity: The days on which the tenant was at home exhibit a distinctive inactivity spike[6] typically beginning shortly after midnight. In the month displayed below, the average time the tenant went to sleep was 0120 (standard deviation $\sigma = 78$ min)

---

[6] In diagrams (n) and (o), the brief sleep interruption at around 0300 was not considered because by looking into the raw data it can clearly be attributed to using the bathroom.

and the average time he got up was 0612 (σ = 46 min), hence the average sleeping time without interruption was 4 h 52 min (σ = 51 min).

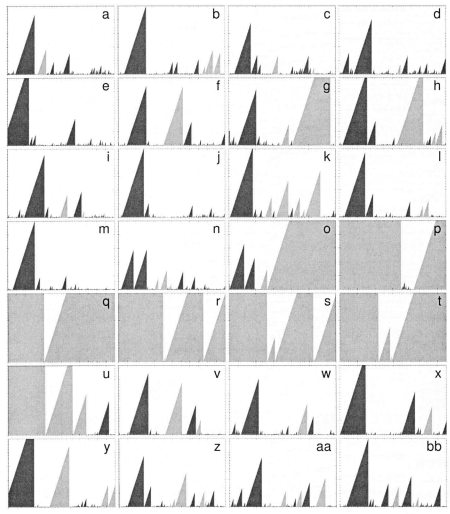

Fig. 6.1: Sample array of 28 single-day inactivity patterns of individual A (time of year: summer[7])[8]

---

[7] Summer and winter are defined as follows: summer = [April, September], winter = [October, March]

[8] All thumbnail images shown here and in the following are equivalent to the one in Fig. 5.6. For the sake of saving space, however, abscissa and ordinate labels have been omitted. The scale of the abscissa is [0000, 2400], the scale of the ordinate is [0 h, 5 h].

The above findings are, however, not yet based on multi-day patterns. Multi-day patterns will be used in the following reasoning steps. Eq. (6.1) defines how maximum-based multi-day inactivity patterns can be calculated.

$$MDI_p(t, S_i^A) := \max_{\forall d \in S_i^A} DI_p(t, d) \qquad (6.1)$$

where $MDI_p(\cdot)$ is the maximum duration of inactivity, S_i is the set of i days from which the maximum-based inactivity pattern is calculated, and A is an arbitrary identifier for the set under consideration.

Using Eq. (6.1), the above series of single-day inactivity profiles can be condensed into a multi-day pattern which only shows the maximum inactivity throughout the days contained in set $S_{28}^{A/June}$. The resulting multi-day pattern is depicted in Fig. 6.2.

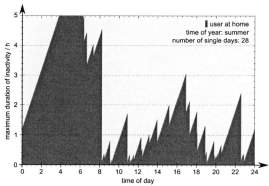

Fig. 6.2: Maximum-based multi-day inactivity pattern: 28 d / summer / individual A (based on Fig. 6.1)

The trend observed in the above series of 28 individual diagrams (Fig. 6.1) is also clearly visible in the highly condensed representation in Fig. 6.2: The tenant typically sleeps between 2300 and 0800. This finding is in agreement with the observations from the individual representations of single days. On the one hand there is, however, a loss of information – the multi-day diagram does not clearly indicate how long the user typically sleeps but only shows a time frame starting at the earliest time the user went to sleep and ending at the latest time he got up. On the other hand, the multi-day diagram clearly shows what durations of inactivity have occurred while the tenant was awake (between 0800 and 2300) which is important to know when setting alarm thresholds. Over a period of 28 days, the user was inactive for three hours once (at around 1700). All other periods of inactivity lasted less than 2.5 hours. Moreover, Fig. 6.2 shows that activity is particularly high (i.e., inactivity is low) in the morning up to 1200 and in the evening hours between 1800 and 2100. How this knowledge will be used for alarm generation will be elucidated in section 6.3.

Another sample set of inactivity patterns recorded in another flat is shown in Fig. 6.3. In contrast to the diagrams shown in Fig. 6.1, the below data exhibits much greater uniformity.

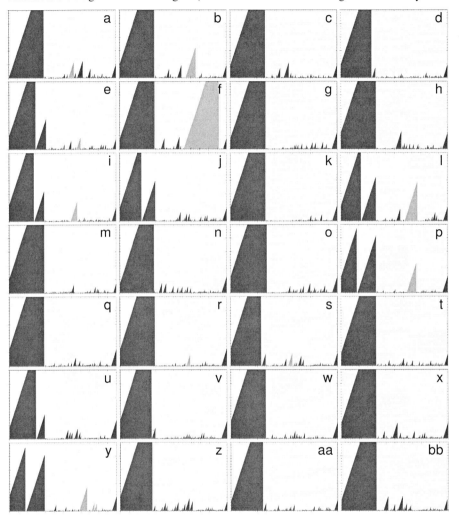

Fig. 6.3: Sample array of 28 single-day inactivity patterns of individual B (time of year: summer)

Tenant B's circadian rhythm is very stable so that recurring features can easily be extracted: The user goes to sleep around 2300 every night and gets up at about 0800 every morning. The reason for the nocturnal inactivity phases being particularly uniform is that the user's bed is outside the range of the presence sensor in the bedroom, i.e., there is no direct line of sight from the sensor to the bed. This is because the sensors were installed prior to the tenants moving in. Thus, in some flats furniture was placed in front of the sensors. Moreover, the user rarely leaves the flat but rather is at home most of the time. In addition, inactivity throughout the day hardly exceeds one hour but is usually even lower.

The above data set of 28 days recorded in the flat of user B is particularly suited for deriving maximum-based multi-day patterns since all days are very similar. Applying Eq. (6.1) to these days yields the multi-day pattern depicted in Fig. 6.4.

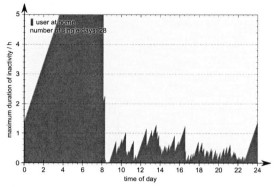

Fig. 6.4: Maximum-based multi-day inactivity pattern: 28 d / summer / individual B (based on Fig. 6.3)

Comparing Fig. 6.2 and Fig. 6.4 corroborates the conclusion that the daily routine of individual B is more uniform than that of individual A. Over a period of four weeks, the longest inactivity phases observed during the awake period amounted to approximately 75 min. Hence, the inactivity patterns of this tenant will be ideal for linear alarm thresholds that will be introduced in section 6.3.2.

In conclusion, maximum-based patterns do well reflect the typical behaviour of users exhibiting rather constant activity and inactivity patterns whereas the maximum inactivity pattern tends to overrepresent inactivity durations of users with vastly varying inactivity patterns as each one-time peak will appear in the long-term pattern.

6.2.3. Mean Value-based Multi-Day Patterns

As shown in the previous section, maximum-based inactivity patterns are a suitable means of condensing multi-day data into a single reference pattern if the individual days' patterns are strongly uniform. However, the resulting multi-day pattern may be rendered unusable if any of the input profiles was very irregular or even faulty. This may be the case if the user's daily routine vastly differs from day to day or if extraordinarily exceptional behaviour was observed on any of the days forming the basis for the calculation. A cause for truly faulty observations may be the malfunctioning of any of the sensors or the communication bus. If, for example, one of the presence sensors fails to signal the start of activity or if the communication channel (KNX bus) is unexpectedly down so that no telegrams can be transmitted, continuous inactivity may be recorded, leading to the observation of seemingly uninterrupted inactivity. Such a long inactivity period will outweigh every other pattern recorded if only the maximum duration of inactivity is considered. Using the same logic, a failure of a sensor or the bus can as well lead to the erroneous recording of apparently continuous activity – this kind of error would, however, have a lesser impact on the calculation of multi-day patterns as they are based on inactivity.

The reason for this kind of failure is that once no more telegrams are received by the data processing unit, no state changes will be triggered any more. If no state changes can be performed, the current states (i.e., activity or inactivity, respectively) will persist until new telegrams will be received again.

The diagrams in Fig. 6.5 shall exemplify such occurrences. Two *distinct* incidences are displayed: first, diagrams (a)-(b) illustrate a data acquisition failure of about 23 hours and second, diagrams (c)-(i) illustrate missing data over a period of approximately 150 hours. Both of these cases did actually occur.

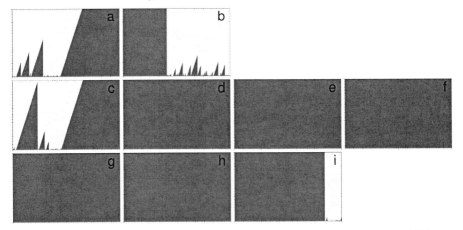

Fig. 6.5: Examples for erroneous data acquisition: (a)-(b) 23 h of missing data; (c)-(i): approx. 150 h of missing data

When calculating maximum-based 28-day patterns of the months in which the above two faulty single-day data sets were observed, the faulty data sets would render the entire multi-day pattern useless. The resulting maximum-based multi-day patterns would show continuous inactivity of extremely long durations so that –in terms of visual representation as introduced above– most of the diagram would be entirely grey because the duration of inactivity exceeds the scale of the diagrams by far. In order to illustrate this effect, maximum-based 28-day diagrams from the months of two distinct data samples from Fig. 6.5 (a)-(b) and (c)-(i), respectively, are shown in Fig. 6.6 (a) and (b), respectively.

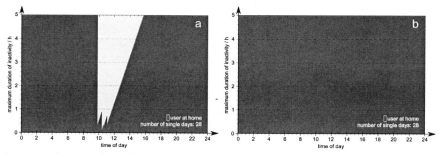

Fig. 6.6: 28-day max-based inactivity patterns resulting from (a) *Fig. 6.5 (a)-(b)* and (b) *Fig. 6.5 (c)-(i)*

The two above maximum-based multi-day patterns indicate that they are not a suitable means for describing the average long-term behaviour of a user if inadvertently recording erroneous data at some stage cannot be ruled out. Thus, a multi-day pattern calculation method needs to be established that is not as susceptible to misrecorded inactivity data as the MDI. This necessity leads to the implementation of mean value-based multi-day patterns described by the average duration of inactivity ADI_p as expressed in Eq. (6.2).

$$ADI_p(t, S_i^A) := \frac{\sum\limits_{d=1}^{i} DI_p(t,d)}{\sum\limits_{d=1}^{i} P(t,d)} \qquad (6.2)$$

$$P(t,d) = \begin{cases} 1 & \text{if tenant at home on day } d \text{ at time } t \\ 0 & \text{elsewhere} \end{cases}$$

where $P(t,d)$ is the number of days on which the tenant was at home at the specific time t.

Other than the maximum-based patterns, the ADI is to some degree resistant to outliers so that non-typical inactivity peaks do not distort the long-term pattern. However, calculating the average duration of inactivity holds the risk of flattening the long-term pattern too strongly. The following examples shall exemplify the method.

Applying Eq. (6.2) to the 28 day-period containing the data from the two days shown in Fig. 6.5 (a)-(b) yields the mean value-based patterns shown in Fig. 6.7. Diagram (a) illustrates that calculating the mean value of activity over a long enough period, i.e., 28 days, can indeed compensate for the effects of short-term misrecordings of inactivity data over up to 24 hours. The overnight failure to record the inactivity data properly is no more directly visible in the aggregate diagram, contrary to the maximum-based multi-day diagram shown in Fig. 6.6 (a) in which the full impact of the misrecorded data comes to the fore. In comparison to Fig. 6.7 (b) for whose calculation the two days on which data acquisition had failed were omitted, (a) shows only slightly higher inactivity throughout the day: Assuming a duration of 23 hours for the erroneously observed inactivity period yields a mean error E of $E(t)$=11.5 h. In turn, dividing this error by $P(t,d)$ (most likely 28 at night when the tenant typically sleeps, i.e., is at home and is inactive; less throughout the day as the tenant sometimes may be active or absent) causes an increase of the "regular" average inactivity ADI_p of approximately 25 minutes due to the spurious inactivity phase. Thus, it can be concluded that if only malfunctions of up to 24 hours are allowed, the maximum impact this can have on the mean value-based multi-day diagram is $24 h / P(t,d) \approx 1 h$ on two conditions: First, it is assumed that the tenant is regularly at home, i.e., $P(t,d) \approx 24$. Second, the maximum error only becomes manifest at the end of such a malfunctioning period, i.e., it is only accumulated at the end of the entire duration of a temporary system failure. Thus, it is believed that faults of up to 24 hours during the learning phase are acceptable. In the present example, the duration of the spurious inactivity period was 23 hours, ending at 1000. The two diagrams in Fig. 6.7 are in perfect accordance with the above reasoning: When comparing (a) and (b), an error of about 1 hour can be seen in (a), most notably at the end of the misrecording period at 1000.

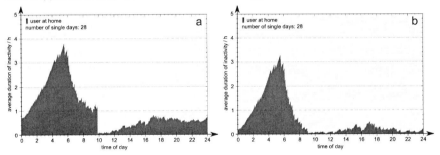

Fig. 6.7: 28-day mean value-based inactivity patterns: (a) including erroneous data from two days
        (Fig. 6.5 (a)-(b)), (b) without that data

When looking at the seven day telegram reception failure (Fig. 6.5, (c)–(i)), however, it turns out that the mean value-based multi-day patterns are by no means a silver bullet for coping with the challenges posed by persistent malfunctions. Fig. 6.8 shows two mean value-based diagrams based on the seven day failure. The ordinate of graph (a) is equivalent to the ones shown in Fig. 6.7 – the standard ordinate scaling of [0, 5] does not nearly suffice to represent the mean average inactivity over a period of 28 days. Only when changing the ordinate scaling to [0, 24], the average inactivity duration becomes apparent: During the night (i.e., $P(t,d) \gg 20$), the average inactivity ADI_p is approximately 15-19 hours, whereas it is 20 and more hours during the day (i.e., $P(t,d) \ll 28$). The reason for this behaviour is that in the unlikely case of the occurrence of inactivities of more than 24 hours, the recurring inactivity is integrated multiple times. In the above example in Fig. 6.5 (c)-(i), the spurious inactivity period started on day (c) at around 1100. On day (d), inactivity lasting 24 hours was observed at 1100, on day (e) inactivity of 48 hours was observed, etc. When calculating $ADI_p(1100, S)$, those seven days alone contribute $24 + 48 + 72 + 96 + 120 + 144 = 504$ h to the inactivity sum and thus –even when letting $P(t,d) = 28$– a minimum of 18 h to the mean inactivity at 1100.

Hence, spurious inactivity of more than 24 hours cannot be compensated for, i.e., removed or sufficiently levelled, using the mean value method.

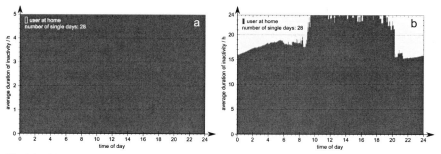

Fig. 6.8: 28-day mean value-based inactivity patterns: (a) including data from seven days (*Fig. 6.5 (c)-(i)*) with standard five hour ordinate scale, (b) with 24 hour ordinate scale

The impact of misrecorded inactivity could be mitigated by increasing the number of days upon which the multi-day patterns are based. Increasing this time span, however, raises several other issues: Too long periods for calculating multi-day patterns might blur trends in the data that might otherwise be detected as long-time changes will be completely incorporated in the reference pattern. Moreover, increasing the set of days used for deriving the multi-day patterns also boosts the probability of observing more than one misrecording whose impacts on the multi-day pattern could then multiply. In general, however, it can be stated that creating multi-day average inactivity patterns is a robust method capable of mostly making up for minor inconsistencies in the data being processed.

ADI can of course also be applied to fault-free data: When calculating the mean value-based multi-day profiles of the two sets of error-free inactivity profiles shown in Fig. 6.1 and Fig. 6.3, the two average inactivity diagrams as shown in Fig. 6.9 result.

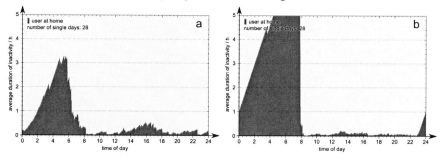

Fig. 6.9: 28-day mean value-based inactivity patterns. (a): based on Fig. 6.1 (individual A), (b): based on Fig. 6.3 (individual B)

The comparison of Fig. 6.2 and Fig. 6.9 (a) shows that the general trends in the daily routine persist regardless of the algorithm used for deriving the multi-day pattern. As mentioned above, tenant A's average sleeping time was 4 h 52 min during the training period. However, since there were significant temporal offsets between the sleeping periods, i.e., they started at considerably varying times and consequently ended at different times, it is not to be expected that the peak of the mean inactivity in Fig. 6.9 (a) reflects this average sleeping time – the peak is found between 0400 and 0530 and reaches a duration of approximately 3 h 15 min. Throughout the day, the general trend exhibited in the maximum-based diagram also persists: The inactivity peak around 1600 in Fig. 6.2 is also clearly visible in the mean value-based diagram.

In case of Fig. 6.4 and Fig. 6.9 (b), similar observation can be made. The nightly sleeping period was very constant throughout the 28-day sample period. Thus, no significant differences between the maximum-based and the mean value-based diagrams can be detected regarding the beginning, duration, and end of the sleeping period. The overall low inactivity throughout the awake period becomes almost unnoticeable in the mean value-based diagram – there is no mean inactivity of more than 10-15 minutes throughout the day.

Both examples illustrate that the distinctive features of the original single-day inactivity patterns are quite resilient and not easily eradicated by condensing the initial information. However, neither the maximum-based nor the mean value-based patterns seem to reflect the users' behaviour appropriately. The maximum-based patterns are easily influenced by outliers and errors in the data acquisition process whereas the mean value-based diagrams tend to smooth the data too substantially so that faulty conclusions might be drawn regarding the "real" inactivity pattern throughout the day. Thus, in the following section, a methodology for removing obvious outliers and only subsequently calculating the condensed patterns will be introduced to merge the advantages of the two above methods.

6.2.4.   Multi-Day Patterns with Outliers Removed

As illustrated above, both the maximum-based and the mean value-based multi-day patterns yield somewhat distorted representations of the long-term behavioural patterns of the user – in the former case, exceptionally long inactivity periods render the multi-day patterns non-representative, in the latter case, the typical daily routine tends to be smoothed too heavily.

This situation is to be mitigated by removing outliers before calculating long-term patterns. In order to make an educated choice for one particular method for removing outliers out of the abundance of outlier tests, the density function of the distribution of inactivity durations throughout a day needs to be examined. All six histograms[9] –representing different times of the day– shown in Fig. 6.10 were thus tested against typical probability distributions often encountered in experimental sample data: normal, uniform, exponential, and double exponential distributions. Neither the chi-square test nor the KOLMOGOROV-SMIRNOV test yielded any accordance of the distributions found in the histograms with the above probability distributions. Hence, well-known outlier tests such as CHAUVENET'S criterion [Taylor, 1996],

---

[9] In this work, all histograms are based on 7 bins since the number of bins should be in accordance with the population size. Having a maximum population size of 28, 7 bins seem reasonable as they provide sufficient resolution while –at the same time– avoid too many empty bins.

PEIRCE'S criterion [Peirce, 1852], or GRUBBS' test for outliers [Grubbs, 1950] cannot be applied since they all require the sample to be normally distributed.

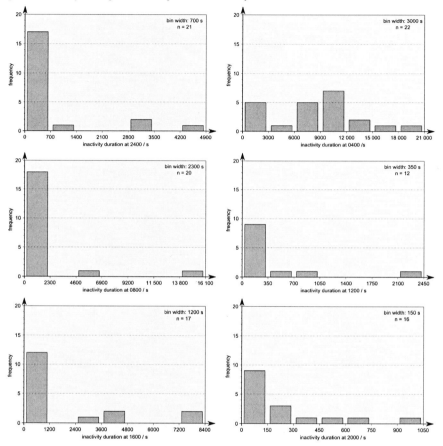

Fig. 6.10: Histograms of the inactivity duration at different times of the day for individual A in a summer month[10]

In order to overcome the problem that the captured data is not governed by an easily identifiable probability distribution, partly due to the small population size, the data is represented using box plots. Box plots are a robust and "resistant" means in descriptive statistics for visualising numeric data and identifying outliers. One of the major advantages of box-plotting over other outlier removal algorithms is that it also works on non-normally distributed data [Brant, 1990]. Early box plots were characterised by five descriptive parameters [McGill et al., 1978]: (1) the smallest observation (sample minimum), (2) the lower quartile *Q1*, (3) the median *Q2*, (4) the upper quartile *Q3*, and (5) the largest observation (sample maximum).

---

[10] n is the number of days on which inactivity had been observed at the times represented by the six histograms.

Later, this initial concept had been expanded to permit the identification of outliers (see Fig. 6.11). The interquartile range *IQR* was introduced by [Tukey, 1990]. Two values for a factor *k*, 1.5 and 3.0, were arbitrarily defined by [Brant, 1990] – however, this choice had been thoroughly tested and the significance been verified so that these values for *k* are now commonly used, thus helping to ensure comparability of results. The factor *k* determines which observed values should be regarded as outliers: Observations within the interval [*Q1*–1.5×*IQR*; *Q3*+1.5×*IQR*] are considered valid. Recorded values falling into either one of the two intervals [*Q1*–1.5×*IQR*; *Q1*–3.0×*IQR*] and [*Q3*+1.5×*IQR*; *Q3*+3.0×*IQR*] are likely to be outliers. Finally, the probability of measurements outside the interval [*Q1*–3.0×*IQR*; *Q3*+3.0×*IQR*] being outliers is statistically significant.

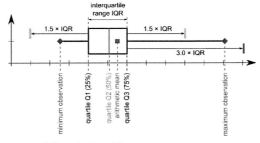

Fig. 6.11: Schematic representation of a box plot

In the above schematic illustration of a data set by means of a box plot, the minimum observation is thus to be considered a valid measurement, whereas the maximum observation might be considered an outlier – depending on the experimental set-up as decided upon by the experimenter. No values were recorded outside 3.0×*IQR* which would have to be considered statistically significant outliers.

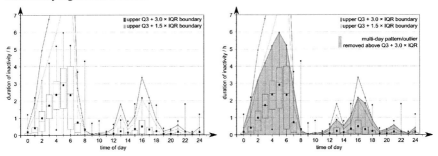

Fig. 6.12: Box plots of the inactivity data from individual A and resulting new multi-day pattern

Applying this methodology to the inactivity data recorded in the flat inhabited by individual A yields the diagram shown in Fig. 6.12. In accordance with the observed differences in the maximum-based and the mean value-based diagrams shown in Fig. 6.2 and Fig. 6.9 (a), particularly noticeable scatter in the observed inactivity durations can be found at night-time (due to different times at which user A goes to sleep) and also in the afternoon from 1500 to 1800. Striking outliers had also been recorded at 2200 and 2400.

When investigating the inactivity durations throughout all day in detail, however, it turns out that sometimes even the largest observed inactivity is still within the ($Q3 + 3.0$ IQR) range while it is well outside this boundary at other times. In the latter case, e.g., at 0800, this indicates a statistically significant outlier. Thus, observations exceeding this boundary (red line in the above diagrams) are discarded. Consequently, at a given time t, the duration of inactivity with outliers having been eliminated DIE_p is either the value of the largest observation at that time (if below the boundary) or equivalent to the boundary (if equal to or greater than the boundary). A formal representation of this rule is given in Eq. (6.3).

$$DIE_p(t, S_i^A) := \begin{cases} MDI_p(t, S_i^A) & \text{if } MDI_p(\cdot) < Q3 + 3.0 \times IQR \\ Q3 + 3.0 \times IQR & \text{elsewhere} \end{cases} \qquad (6.3)$$

The resulting DIE_p graph depicted in Fig. 6.12 is thus an intermediate between the maximum-based and the mean value-based diagrams shown above. It is believed that this diagram is a good means of representing the typical, average user behaviour over long periods of time. At night-time, the long-time inactivity curve is not flattened because it exhibits considerable variance and long periods of sleep are very likely to occur. During day-time, however, the variance of the data is much smaller and several data points are thus discarded.

In Fig. 6.13, the fully detailed multi-day inactivity pattern DIE_p resulting from the 28 single-day data sets of individual A (see Fig. 6.1) is presented. Fig. (a) shows the resulting inactivity pattern without outliers. For the convenience of comparison, Fig. (b) shows the same pattern along with two overlays, i.e., the maximum- and the average-based pattern. It is clearly visible that several peaks found in the maximum-based multi-day pattern have disappeared. The three most notable peaks that cannot be found in the outlier-free multi-day pattern are those at 0800, 1100, and 2230. At the same time, the diagram also shows that no overly greedy discarding of inactivity occurs that could be observed in case of the mean value-based long-term pattern. Especially at night-time, no significant discarding of observations occurred, hence ensuring that normal sleep patterns of up to 6 hours will not trigger alarms later on when inactivity thresholds will be applied as alarm criteria. Thus, it can be concluded that discarding outliers using the box-plot-criteria is a viable means of identifying typical long-term behaviour without admitting spurious inactivity periods into the multi-day reference patterns.

Using the same logic, outlier-free long-term patterns can be calculated for other users as well. In the following, the data of individual B (see Fig. 6.3) will be processed accordingly. Fig. 6.14 shows the histograms of the data recorded in flat B. As already mentioned above, the daily rhythm of user B exhibits a far greater uniformity than that of user A. This is clearly reflected in the skewness of the below histograms. Apart from the histogram showing the observations made at 2400, all other histograms show a distinct tendency towards either mostly short periods of inactivity (0800, 1200, 1600, and 2000) or mostly long inactivities (0400). This finding is in perfect accordance with the diagrams showing the inactivity patterns of the single days: During the day, only very short periods of inactivity were observed. In addition, user B goes to sleep at a clearly defined time so that at around 2400 approximately the first hour of sleep has usually elapsed. Four hours later, at 0400, the average duration of inactivity is thus about five hours which matches the data in the 0400 histogram.

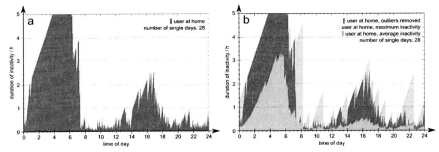

Fig. 6.13: (a) 28 day pattern of individual A with outliers removed; (b) in comparison with maximum- and average-based patterns

Moreover, it needs to be noted that the bin width during day-time is much smaller for user B than for user A. This again underscores that the overall activity of user B is higher *or* that user B triggers the sensors more frequently by exercising the same activity, e.g., resides closer to the motion detectors so that they fire more easily.

However, because of the greater uniformity of the daily rhythm, the difference between the outlier-free long-term pattern (Fig. 6.15) is not expected to differ as much from the maximum-based one as in case of user A.

Fig. 6.15 shows that the outlier-free pattern is closer to the maximum-based pattern than to the mean value-based one. Due to the overall low inactivity throughout the day, fewer extreme outliers had to be removed. At night, however, the maximum-based, the mean value-based, and the outlier-free long-term patterns exhibit great similarity. This is due to the very constant sleep pattern of user B.

To conclude, the outlier-free multi-day patterns will be used as basic reference patterns for most subsequent steps of data interpretation and alarming steps. Since the graph is rather jagged, however, it will first be subjected to an exponential smoothing algorithm to flatten the most prominent spikes.

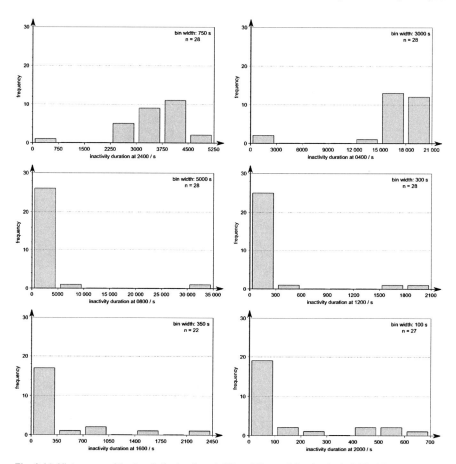

Fig. 6.14: Histograms of the inactivity duration at different times of the day for individual B in a summer month

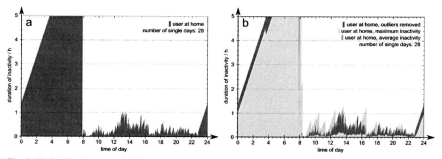

Fig. 6.15: (a) 28 day pattern of individual B with outliers removed; (b) in comparison with maximum- and average-based patterns

6.2.5. Smoothing of Multi-Day Patterns

As shown in the previous section, long-term inactivity patterns with potential outliers being removed promise to be the most suitable reference pattern for analysing inactivity and, based thereupon, identifying potentially dangerous situations and raising alarms. The resulting long-term patterns are, however, rather jagged. Thus, the jagged long-term patterns will be exponentially smoothed as it may be detrimental for some of the data processing and analysing steps introduced in the following chapters to work on discontinuous data.

A plenitude of exponential smoothing algorithms has been published. A comprehensive overview is given in [Gardner, 1985]. Since the objective of smoothing the outlier-free multi-day patterns is reducing the noise that resulted from discarding data points considered outliers, smoothing is only to be performed locally, i.e., not affecting the graph over long periods of time. The overall trend exhibited by the multi-day pattern should not be changed. Thus, simple, non-seasonal smoothing promises to be a suitable means of removing sudden peaks or drops in the long-term inactivity pattern. This exponential smoothing process is essentially a weighted moving average calculated for each second of the day, partially taking into account previous data points. According to [Gardner, 1985], simple smoothing is optimal for large sample sizes which is assumed to be the case considering 86 400 data points (i.e., seconds in a day). Smoothing the outlier-free data is performed using Eq. (6.4):

$$DIE_p^*(t, S_i^A) := \alpha \times DIE_p(t, S_i^A) + (1-\alpha) \times DIE_p^*(t-1s, S_i^A) \qquad (6.4)$$

where α is the smoothing factor and $(1-\alpha)$ is the discount factor.

The larger the α value is, the faster adapts the smoothed graph to the source graph. In contrast, a large discount factor means that substantial influence of historic data on the current smoothed value persists for many iteration steps. Fig. 6.16 illustrates the correlation of the smoothing factor, the elapsed time, and the influence a historic data point still has on the current smoothed value.

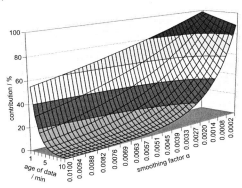

Fig. 6.16: Contribution of historic data points to the current, smoothed graph

For this work, the α value was chosen as 0.0060. Setting the smoothing factor to 0.0060 means that already after a short time historic data, i.e., the values of the smoothed curve from a few minutes ago, hardly contribute to the current value of the smoothed curve. Values one minute old are taken into consideration by 70%, those 5 minutes old contribute 16%, and those 10 minutes old contribute 3% to the currently computed value. The actual choice for the smoothing factor is based on the (arbitrary) reasoning that the primary objective of the additional smoothing step is removing the noise incurred by discarding statistically significant outliers in the multi-day patterns. The second requirement is that the smoothed graph shall only differ as little as possible from the original one. Setting α to 0.0060 fulfils these two demands: The noise can be reduced significantly while, at the same time, the mean difference $\overline{\Delta_{\mathrm{DIE}}}$ of $DIE_\mathrm{p}(\cdot)$ and $DIE_\mathrm{p}^*(\cdot)$ is smaller than 0.1 second. Even though the individual $\Delta_{\mathrm{DIE}}(t)$ can range from more than -1 hour to more than $+1$ hour (because of the lag of the smoothed curve when the original graph exhibits discontinuities), it can thus be assumed that the smoothed curve is well-suited for ultimately representing the long-term inactivity patterns of the users. In Fig. 6.17, the original outlier-free graph (a) and the smoothed curve (b), captured in flat A, are placed in juxtaposition. The two diagrams illustrate that the overall inactivity curve progression remains unaltered. The high-frequency noise, particularly pronounced during phases of low inactivity, is mostly eliminated.

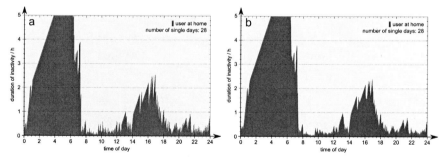

Fig. 6.17: Flat A: (a) Unprocessed outlier-free multi-day inactivity pattern; (b) same pattern smoothed

Long-term, multi-day inactivity patterns of the type shown in Fig. 6.17 (b), i.e., with outliers removed and subsequently smoothed, will be used as reference patterns in the following unless stated otherwise.

6.3.   Thresholds

6.3.1.   General Concept of Thresholds

Miscellaneous alarm thresholds, derived from multi-day patterns introduced above, are the principal instruments for identifying potentially dangerous situations. Based on such thresholds, alarms shall be triggered if any inactivity longer than allowed by the threshold is observed. The simplest threshold one can use for generating inactivity alarms is a static linear alarm threshold $SLAT_{static}(t) = x$ minutes = const. Once that static, linear threshold is exceeded by the current inactivity duration at any given time, an alarm will be triggered. The functional principle of static thresholds and more sophisticated threshold types will be discussed in detail in the following sections.

6.3.2.   Static Linear Alarm Thresholds

The underlying principle of static linear alarm thresholds (SLATs) is rather straightforward: A fixed time limit is defined and an alarm is triggered upon the current duration of inactivity exceeding this threshold at any point in time. Such an alarm threshold is easy to implement and easy to explain to the users. However, SLATs are expected to generate significant numbers of false alarms since they cannot accommodate inactivity minima or maxima in the course of a day or in a multi-day pattern. This is why the concept of SLATs will be developed further into mSLATs –masked SLATs– only active during certain times of the day. By masking for example night-time, false alarms arising from extended inactivity periods while the user is asleep can be avoided.

SLATs and mSLATs will be discussed in detail in the following. Fig. 6.18 shows sample SLATs applied to a single-day inactivity graph from flat A.

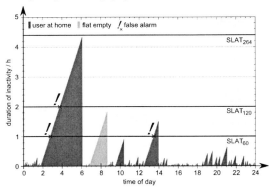

Fig. 6.18: Sample SLATs applied to single-day data from flat A

The basic alarm principle can be expressed formally as shown in Definition 6.1. In this definition, *further steps* are not yet precisely specified as section 7.2: Alarm Handling and

Forwarding will deal with all necessary measures to be taken in order to react to a possible health issue.

> **Definition 6.1:** Alarm principle of SLATs
>
> **IF**   (inactivity $DI_p(t, d)$ exceeds alarm threshold $SLAT_x$)
> **THEN**   a potential health threat is to be assumed **AND**
>   further steps need to be taken

The above example, however, illustrates that a *static* threshold active 24/7 would have to be very high, i.e., several hours. In the following Fig. 6.19, any 24/7 SLAT of less than six hours would trigger at least on false alarm on the particular day shown in the diagram although inactivity never exceeded half an hour during day-time.

However, it must be acknowledged that even response times of several hours constitute a major advancement compared to the status quo of (manual) emergency detection in homes without any AAL systems which heavily relies on third persons who by chance come to visit the senior person in need for help. In the following, more advanced alarming schemes will be introduced that allow lowering response times to two or three hours. Using any of these thresholds, it can effectively be prevented that senior citizens experience a case of emergency in their homes and have to wait for days or even longer for help to arrive. Ultimately, such AAL systems can save lives since even seemingly harmless incidents can turn out to be fatal if they cannot be discovered within reasonable time.

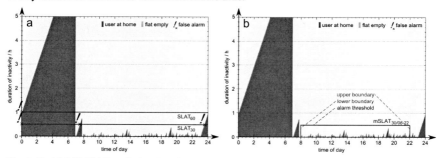

Fig. 6.19: (a) 24/7 SLAT applied to data from flat B; (b) masked mSLAT applied to same day

One conceivable remedy for the problems arising when having to deal with long nightly periods of inactivity is *masking* the SLAT Fig. 6.19 (a), thus yielding an mSLAT (b) (see Definition 6.2). The lower and upper boundary (*LB, UB*) of the mSLAT have to be defined according to the sleeping pattern of the respective person being monitored. Thus, the above alarm rule for SLATs has to be expanded to accommodate the period of operation (*PO*) constituted by the lower and upper boundary of the mSLAT, i.e., $PO = [LB, UB]$. Within *PO*, the alarm threshold *ALT* is constant.

> **Definition 6.2:** Alarm principle of mSLATs
>
> **IF**   (*time of day t is outside PO*)
>
> **THEN** automatic alarming is disabled
>
> **ELSE**   **IF**   (*inactivity $DI_p(t, d)$ exceeds alarm threshold $mSLAT_x$*)
>
>   **THEN**   a potential health threat is to be assumed **AND**
>
>   further steps need to be taken

In case of the user living in flat B, 0800 and 2200 seem a reasonable choice for *LB* and *UB*, respectively. As shown in Fig. 6.19 (b), no false alarms would have been triggered if a masked alarm threshold *ALT* of 30 minutes, starting at 0800 and ending at 2200, i.e., $mSLAT_{30/08-22}$, would have been set. The above inactivity pattern, however, represents only a single day's inactivity. Thus, general statements about which mSLAT may be appropriate as a criterion to raise alarms in case of suspected cases of emergency cannot be made.

Considerably more information about long-term inactivity patterns can be extracted from the multi-day graphs introduced above. In case of the maximum-based multi-day patterns, setting the mSLAT higher than the highest peak will eliminate most false alarms, even though on some future days the observed inactivity may be longer than in the maximum-based diagram which is based on a monitoring period of 28 days. Thus, occasional false alarms may still occur. Fig. 6.20 illustrates how the mSLATs can be set in flat A and flat B, respectively.

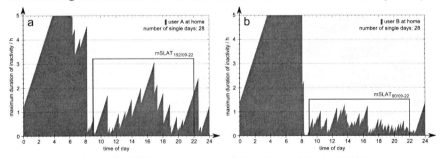

Fig. 6.20: Dependence of mSLATs on maximum inactivity patterns: (a) flat A; (b) flat B

Comparing the two above diagrams and their respective mSLATs clearly shows that mSLATs can be useful in some instances while they may be less helpful in other cases. In diagram (a), showing a 28 day maximum-based pattern from user A, an mSLAT of 192 minutes would have been the obvious choice if false alarms in the considered 28 day period had been to be avoided. As explained above, however, a threshold of several hours is not acceptable in case of a genuine emergency. Assistance has to be provided much quicker. Even in case of a non-life-threatening fall, having to wait for more than three hours until help will arrive is by far too long – during day-time, only response times of up to a maximum of two hours, in exceptional cases of up to three hours, are deemed acceptable. Thus, in diagram (b) an mSLAT of only 90 minutes would have to be considered a reasonable choice. During a period of 28 days, no inactivity periods of more than about 75 minutes had been observed.

Other than in case of $mSLAT_{192/09\text{-}22}$ in diagram (a), 90 minutes appear to be an acceptable threshold. After the occurrence of a fall, faint, etc., such a period until the response time has elapsed can most likely be endured by an affected person.

The following diagrams and tables evaluate the suitability of the mSLATs as defined above. In Fig. 6.21, the number of false alarms in the sample flats A and B triggered by the above two mSLATs are given. In addition, the two mSLATs having been defined based on Fig. 6.20 were each complemented with a second set of false alarm data in order to illustrate how a change to the chosen mSLAT affects the number of false alarms generated.

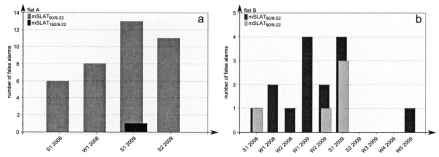

Fig. 6.21: False alarms in flat A and flat B in several months based on mSLATs[11]

In case of flat A (diagram (a)), $mSLAT_{192/09\text{-}22}$ yields only one false alarm in four months. It needs to be noted, though, that 192 minutes are a rather high threshold not very suitable for real application (see above). However, when changing the threshold to a more responsive $mSLAT_{90/09\text{-}22}$ to avoid long response times, the number of false alarms in those four months increases drastically, thus rendering a 90 minutes threshold useless for alarming. Up to 14 false alarms per month would result from this $mSLAT_{90/09\text{-}22}$. This is not acceptable either as it would mean approximately one false alarm every other day.

When looking at flat B (diagram (b)), it turns out that the $mSLAT_{90/09\text{-}22}$ seems to be well suitable as threshold for raising alarms in that flat. Applying $mSLAT_{90/09\text{-}22}$ yields zero, one, or –however, in only one case– three false alarms per month which is deemed acceptable. For demonstration purposes, the number of false alarms generated when using $mSLAT_{90/08\text{-}22}$ instead is also displayed. The single-day diagrams from all months considered in the above diagram (b) show that the tenant does usually get up before 0800. In a few exceptional cases, however, the tenant got up later than at 0800. Thus, false alarms are raised on days on which the person is still sleeping at 0800. As expected, the number of false alarms for $mSLAT_{90/08\text{-}22}$ has thus risen compared to $mSLAT_{90/09\text{-}22}$, but the increase was only moderate (up to a total of four false alarms per month). Hence, this example illustrates the trade-offs quite well that

---

[11] To protect the privacy of the tenants, the actual months will not be revealed. Instead, summer months are labelled Sn and winter months Wn. In addition, the months that are suitable for this evaluation are not contiguous because only months in which data from all days are available will be considered to avoid biased results. Since the tenants can switch off the control unit at their will, however, complete data is only available for a limited number of months in a year.

have to be made between a wide period of operation and quick response times on the one hand and low number of false alarms on the other hand.

In conclusion, it can be said that mSLATs can indeed be used as a basic alarming criterion if the overall inactivity level throughout the day is relatively low (e.g., in flat B) whereas they may not be ideally suited in other cases in which long inactivity periods are likely to be encountered (e.g., in flat A). Finally, a more detailed breakdown of the causes of the false alarms in flat A and B is given in Table 6.1 and Table 6.2.

Table 6.1: Number of false alarms in flat A in different months applying two mSLATs

	Flat A							
	$mSLAT_{90/09\text{-}22}$				$mSLAT_{192/09\text{-}22}$			
Month	**total false alarms**[12]	tenant slept longer than usual	failure of monitoring system	**genuine false alarms**	**total false alarms**	tenant slept longer than usual	failure of monitoring system	**genuine false alarms**
S1 08	6	–	–	6	0	–	–	–
W1 08	8	–	–	8	0	–	–	–
S1 09	13 (2 × 2)	1	–	12	1	–	–	1
S2 09	11	–	–	11	0	–	–	–

The above table illustrates that in flat A a large number of false alarms is to be expected when applying an mSLAT with $ALT = 90$ minutes. Even though the single-day inactivity patterns did not suggest this threshold, it has been investigated for reasons of comparability and completeness. Even when discarding those false alarms that were due to the tenant lying in, between 6 and 12 genuine false alarms remain per month. When applying an ALT of 192 minutes, as the multi-day pattern (Fig. 6.20 (a)) suggests, only a single false alarm in four months is encountered. This result supports the assumption that mSLATs represent the user behaviour quite well if they are based on empirical evidence, i.e., long-term patterns.

The below table showing the numbers of false alarms from flat B corroborates the above findings. Using $mSLAT_{90/09\text{-}22}$ as indicated in Fig. 6.20, only five false alarms in ten months would have occurred. This number of false alarms is assumed to be acceptable for most users. When expanding PO to 0800–2200, several more false alarms would have been triggered. Closer inspection of the single-day patterns shows that the tenant usually gets up shortly before 0800 but sometimes stays in bed slightly longer. Even though the tenant does get up by 0810 at the latest, an alarm is raised anyway. The captured raw data from flat B also shows a failure of the monitoring system. Over a period of approximately three days, no telegrams were received, thus leading to an inactivity phase of three days. If the alarming system had

---

[12] ($n \times 2$) means that on n days two false alarms occurred in a single day.

been armed, a false alarm would have been triggered after 90 minutes and the malfunctioning of the monitoring system would have been noticed and could have been fixed immediately.

Table 6.2: Number of false alarms in flat B in different months applying two mSLATs

Flat B								
	$mSLAT_{90/08-22}$				$mSLAT_{90/09-22}$			
Month	total false alarms	tenant slept longer than usual	failure of moni- toring system	genuine false alarms	total false alarms	tenant slept longer than usual	failure of moni- toring system	genuine false alarms
S1 08	1	–	–	1	1	–	–	1
W1 08	2	2	–	–	0	–	–	–
W2 08	1	1	–	–	0	–	–	–
W1 09	4	4	–	–	0	–	–	–
W2 09	2	1	–	1	1	–	–	1
S1 09	4	1	1	2	3	–	1	2
S2 09	0	–	–	–	0	–	–	–
W3 09	0	–	–	–	0	–	–	–
W4 09	0	–	–	–	0	–	–	–
W5 09	1	1	–	–	0	–	–	–

In summary, the above tables illustrate that mSLATs exhibit similar shortcomings as SLATs when trying to avoid false alarms even though they eliminate the problems arising at night-time when the tenants are asleep and long periods of inactivity can occur. Due to possible high inactivity peaks at random times of the day –which may be perfectly legitimate– employing (m)SLATs while at the same time trying to avoid false alarms altogether is no viable way of detecting potentially dangerous situations the user may be in. Two countermeasures are conceivable to mitigate this deficiency: Either, non-static thresholds are to be applied that are able to accommodate different levels of inactivity during the course of the day, or a certain number of false alarms is to be allowed. In the former case, those non-static thresholds could, for instance, be based upon the outlier-free multi-day patterns described above. This approach will be elucidated in detail in the following section. The latter measure –tolerating false alarms under certain conditions– turned out to be a very useful means of raising and handling alarms. This entire concept will be introduced in chapter 7: Alarm Management.

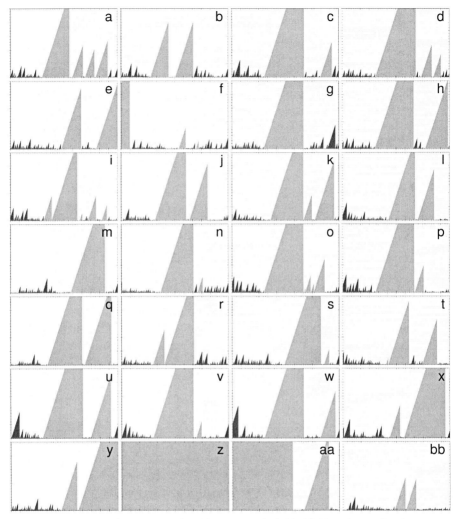

Fig. 6.22: Sample array of 28 single-day inactivity patterns of individual C (time of year: winter)

Last but not least, it needs to be mentioned that there may be sensor array setups in which the motion detector in the bedroom is able to detect activity while the tenant is asleep – for this, the line-of-sight from the motion detector to the bed may not be obstructed as it is the case in flat A and B. In the following, sample data from a flat C in which the line-of-sight to the bed is not obstructed is discussed[13]. Above Fig. 6.22 clearly shows that the nightly sleep

---

[13] The data captured in flat C is deemed to be of very good quality. However, only a very limited amount of continuous data is available. Notwithstanding this issue, data from flat C will be used here for demonstration purposes.

pattern of user C is fundamentally different from those found in flats A and B. The overall inactivity level is very low. Only in few exceptional cases, inactivity periods of more than one hour can be found, and even that only at night-time.

The resulting 28 day maximum-based long-term pattern is shown in Fig. 6.23. As mentioned above, no inactivity exceeding one hour had been observed during the day, and at night-time, the longest periods of inactivity were recorded between 2200 and 0200, however, hardly exceeding two hours. Hence, the data captured in flat C differs from the data gathered in flats A and B in that it exhibits the lowest inactivity both during the day and at night. In the latter case, this is due to the motion detectors being able to detect motion while the tenant is asleep. However, even using the data from flat C, it is not recommended to apply a simple SLAT as this would require a minimum ALT of 2.4 hours, which is not advisable as explained above.

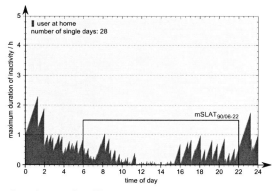

Fig. 6.23: Maximum-based pattern from Fig. 6.22, $mSLAT_{90/06-22}$ (flat C)

Due to the overall low inactivity in flat C, however, an mSLAT is believed to be a viable means of raising alarms in this particular case, as illustrated by the mSLAT mark. Without creating an overabundance of false alarms, the lower boundary of the mSLAT can be shifted to 0600 so that PO covers a larger portion of the day than in flats A and B. Based on the above diagram, even shifting LB to 0200 could be justified, but might in fact lead to undesired false alarms in the middle of the night if the tenant happened to sleep very statically one night. Hence, applying $mSLAT_{90/06-22}$ as an alarm threshold seems reasonable and means that during day-time, the worst case would be having to wait for 90 minutes for help to arrive. At the same time, a time buffer of 30 minutes on top of the longest period of inactivity is provided, thus reducing the likelihood of false alarms significantly.

In Fig. 6.24, the fitness of the chosen mSLAT, based on Fig. 6.23, is demonstrated using data from a winter month, i.e., the month following the one used for establishing $mSLAT_{90/06-22}$. On only one day, $mSLAT_{90/06-22}$ would have caused a false alarm (b). This is assumed to be acceptable. A detailed discussion of why and how false alarms are important and useful will follow in section 7.3. For demonstration purposes, the number of false alarms that would have been triggered using an alarm threshold of 80 minutes (dashed line in the diagrams) has been identified. Eighty minutes is deemed to be among the lowest possible alarm thresholds that will not lead to an unacceptable abundance of false alarms. In the above example, a threshold of 80 minutes would have triggered three false alarms (a-c).

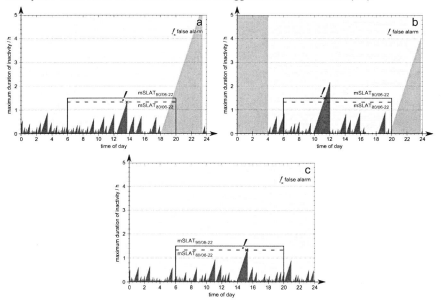

Fig. 6.24: False alarms triggered in flat C by $mSLAT_{90/06-22}$ and $mSLAT_{80/06-22}$, respectively, in a winter month

The findings discussed in this section show that mSLATs can be suitable as basic alarming criteria if the overall inactivity in a flat is reasonably low, i.e., usually less than 60-90 minutes during day-time. This was the case in flat B and C. It can thus be assumed that emergencies of class B can well be detected within a reasonable period of time if prospective users exhibit similar behavioural patterns as those in flats B and C. Nevertheless, the usability of mSLATs as a general alarming criterion is not universal since they can only be applied to that specific type of user behaviour.

6.3.3. Multi-Day Inactivity-derived Thresholds

There are two weak points attached to the basic (m)SLAT approach. First, it needs to be determined by whom the (m)SLAT is to be established. This is an essential question since individual adjustments taking into account user-specific circumstances are both necessary and desirable. Due to vastly varying daily rhythms and sleeps patterns of the users, a "one-fits-all" approach will not suffice. However, since self-determination is one of the fundamental pre-requisites of the entire project, it is not acceptable to have third parties, e.g., family members or medical professionals, define the threshold. Thus, the individual user has to be given the possibility to set their preferred alarm threshold and change it whenever desired.

The second question that needs to be answered is based on what information or criterion the user should make an educated choice for the threshold. If the threshold is set too low, numerous false alarms will result therefrom. In the diagram shown in Fig. 6.18, the 60-minute threshold $SLAT_{60}$ illustrates that two false alarms –at 0230 and 1300– would have been raised if this threshold had been chosen on that particular day shown in the diagram. Even though the necessity of and handling of false alarms will be discussed in detail below (see section 7.3), two false alarms in one day are most likely not acceptable for most users. Even when doubling the alarm threshold ALT to 120 minutes ($SLAT_{120}$), the false alarm at 0330 would still be triggered. The nightly false alarm would be a particular nuisance for the user because the alarm or –if the user fails to cancel the alarm– the steps taken to react to it would certainly annoy the user. In order to avoid false alarms in the example shown in Fig. 6.18, an ALT of more than the highest inactivity peak, e.g., $SLAT_{264}$, would have to be chosen. Doing so, however, would take the entire idea of alarming in case of emergencies to the point of absurdity – if the monitoring system is to detect potential emergencies promptly, an alarm threshold of more than four hours is not useful.

Multi-day inactivity-derived thresholds (MITs) were developed to mitigate these problems encountered when using (m)SLATs. As mentioned above, those are namely issues related to significant differences between the inactivity patterns of day-time and night-time on the one hand and –even more importantly– long inactivity periods of several hours during day-time on the other hand. Coping with long or very irregular inactivity periods requires extremely high alarm thresholds in order to avoid numerous false alarms, thus leading to unnecessarily long times a person would have to wait for help in case of a genuine emergency.

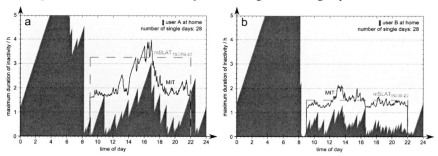

Fig. 6.25: Multi-day inactivity-derived thresholds : (a) flat A, (b) flat B, both with PO = [0900,2200]

In order to circumvent those problems related to (m)SLATS, MITs are based on outlier-free, multi-day inactivity patterns (OMIPs). Since OMIPs are non-linear and take into account typical user behaviour throughout the day, they are considerably lower than SLATs during times of the day when usually low inactivity is observed and are thus deemed to be a suitable base for raising alarms. It is, however, not possible to use an OMIP directly as an MIT. When computing an OMIP (see Eq. (6.4)), all outliers are discarded. Thus, an OMIP only represents average user behaviour which was stripped of exceptionally long periods of inactivity. In order to be used as an MIT, an OMIP needs to be shifted along the ordinate, i.e., extra time needs to be added so that it can accommodate long but still probably normal periods of inactivity that only occur occasionally[14]. Fig. 6.25 exemplifies the functional principle of MITs and how they are established.

The OMIP that belongs to the desired PO is calculated and shifted upwards along the ordinate in such a way that all inactivity peaks –including those that were discarded as outliers– are below the OMIP. The such shifted OMIP becomes the new MIT and can henceforth be employed as an alarm threshold – alarms will be raised if any actual period of inactivity exceeds MIT at the respective time of the day. This process can be formalised as follows (Eq. (6.5)):

$$\Delta_{max} = \max_{\forall t \in PO} \left\{ MDI_p(t, S_i^A) - DIE_p^*(t, S_i^A) \right\}$$

$$\forall t \in PO : MIT_{PO}(t, S_i^A) = DIE_p(\cdot) + \Delta_{max}$$

(6.5)

The resulting alarm rule for the use with MITs is given in the following Definition 6.3. In principle it is the same procedure as in case of the (m)SLATs but the threshold is replaced by an MIT.

Definition 6.3: Alarm principle of MITs		
IF	*(time of day t is outside PO)*	
THEN	automatic alarming is disabled	
ELSE	**IF**	*(inactivity DI_p(t, d) exceeds alarm threshold MIT_{PO}(·))*
	THEN	a potential health threat is to be assumed **AND** further steps need to be taken

It needs to be noted, however, that in the following multiple false alarms that were triggered by a single inactivity period were aggregated into a single alarm. This is due to some sections of $MIT_x(\cdot)$ being quite jagged. The following figure illustrates this effect.

Since all methods and algorithms introduced above are based on historic data captured in various flats, alarms were not raised in real time but simulations were conducted based on the available data. Thus, all inactivity peaks had a given duration and were used for simulating alarms. Therefore, as shown below, a single inactivity period could trigger multiple alarms when the MIT was jagged and the inactivity exceeded the threshold several times (see **1** and **2** in Fig. 6.26). In such a case, all alarms caused by one inactivity spike were aggregated into one. On a side note, the above example also illustrates that due to the jaggedness of the MIT, the maximum allowed duration of inactivity in theory (**3**) cannot always be reached at certain times in practice because the increasing inactivity, having a slope of 1, would have triggered an alarm (**1**) before reaching the maximum (**3**).

---

[14] The aforementioned concerns regarding self-determination of the user are met in that the users are given another instrument –the mean time between false alarms MTFA– to maintain their self-determination in setting their preferred ALT. The concept of the MTFA will be introduced below in section 7.4.

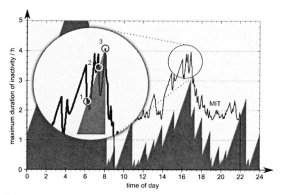

Fig. 6.26: Multiple false alarms caused by a single period of inactivity

Table 6.3: Number of false alarms in flats A and B in different months based on the $MIT(\cdot)$[15]

	$MIT_{09\text{-}22}(\cdot)$ in flat A					$MIT_{09\text{-}22}(\cdot)$ in flat B			
Month	total false alarms	tenant slept longer than usual	failure of monitoring system	genuine false alarms	Month	total false alarms	tenant slept longer than usual	failure of monitoring system	genuine false alarms
S1 08	**1** (0)	– (–)	– (–)	1 (–)	S1 08	1 (1)	– (–)	– (–)	1 (1)
W1 08	**2** (0)	– (–)	– (–)	2 (–)	W1 08	0 (0)	– (–)	– (–)	– (–)
S1 09	**3** (1 × 2) (1)	1 (–)	– (–)	2 (1)	W2 08	0 (0)	– (–)	– (–)	– (–)
S2 09	**2** (0)	– (–)	– (–)	2 (–)	W1 09	0 (0)	– (–)	– (–)	– (–)
					W2 09	0 (1)	– (–)	– (–)	– (1)
					S1 09	2 (3)	– (–)	1 (1)	1 (2)
					S2 09	0 (0)	– (–)	– (–)	– (–)
					W3 09	0 (0)	– (–)	– (–)	– (–)
					W4 09	0 (0)	– (–)	– (–)	– (–)
					W5 09	0 (0)	– (–)	– (–)	– (–)

---

[15] Numbers given in parentheses: false alarms triggered by mSLATS as given above.

The MIT alarm principle was applied to the available recorded data from flat A and B in the same way as the mSLATs. Table 6.3 shows the number of false alarms that would have been triggered if this alarm principle had been active in the respective flats in these months. Results based on $MIT_{09-22}(\cdot)$ are displayed in black whereas the numbers of false alarms triggered by $mSLAT_{192/09-22}$ (flat A) and $mSLAT_{90/09-22}$ (flat B), respectively, are given in parentheses for reasons of comparability. As the above table and Fig. 6.27 illustrate, the number of false alarms in flat A rose by one or two per month.

The overall improvement, however, is substantial against the background that in case of an emergency the response time could be reduced from $mSLAT_{192/09-22}$ to an average response time of 134 minutes, i.e. a reduction of one hour, with the minimum and maximum response times being 96 and 238 minutes, respectively. The average response time ART is determined as follows (6.6):

$$ART(S_i^A, LB, UB) = \frac{\sum_{t=LB}^{UB} MIT(t, S_i^A)}{UB - LB} \qquad (6.6)$$

It can thus be concluded that by tolerating one or two additional false alarms per month, the response time could be lowered by one hour.

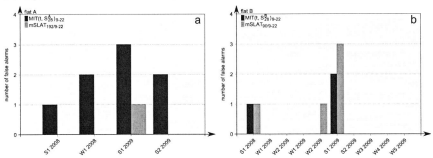

Fig. 6.27: Number of false alarms depending on type of alarm threshold: (a) flat A, (b) flat B

In flat B, the situation is slightly different. Even though the reduction of the response time was not as significant as in flat A (90 minutes using $mSLAT_{90/09-22}$ vs. 87 minutes on average using $MIT_{09-22}(\cdot)$, 71 minutes minimum, 128 minutes maximum), the number of false alarms had been reduced. In ten months, only three false alarms would have been raised using this threshold.

Since MITs are based on the user's typical behaviour, they also represent the inactivity pattern while the user is asleep. Thus, the performance of MITs when being applied 24 hours was also evaluated. Initially it had been assumed that MITs are well suited as alarming criteria

24/7 but it turned out that due to having discarded outliers in the OMIP, especially the transition from the sleeping phase to the wake-up phase is very problematic (Fig. 6.28).

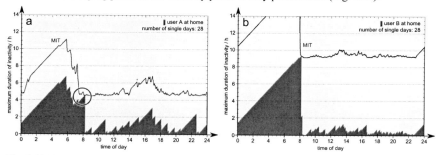

Fig. 6.28: Multi-day inactivity-derived thresholds : (a) flat A, (b) flat B, both with PO = [0000,2400]

The above diagrams clearly demonstrate that the OMIPs have to be shifted along the ordinate to very long response times in order to result in MITs being able to accommodate long inactivity periods caused by the user sleeping longer than normal. Fig. 6.28 (a) illustrates this effect very well: The MIT is established by shifting the OMIP along the ordinate. In the OMIP, however, unusually long periods of inactivity were discarded if having been considered outliers (see circular marker). Thus, the sleep phases ending later than normal in the morning are no more represented in the OMIP. If, however, reasonably low MITs are desired, i.e., similar to $MIT_{09-22}(\cdot)$, all single-day sleep patterns whose wake-up phases were discarded as outliers when establishing the OMIP will now trigger false alarms. Avoiding such false alarms in the morning is only possible by shifting the MIT to response times high enough not to be reached by any matutinal inactivity sleeping phases, as shown in the above figures. In case of flat A in which the tenant exhibits a somewhat irregular sleeping pattern with inactivity peaks of at most six hours, the OMIP would have to be shifted along the ordinate by about five hours. In flat B in which the sleeping pattern is very regular and nightly sleep phases last about nine hours, the OMIP would have to be shifted by almost 10 hours, hence rendering it useless for emergency alarming.

It can thus be concluded that an MIT_{00-24} is no viable alarm threshold for raising alarms based upon the inactivity of the tenants unless false alarms are tolerated already during the process of computing the alarm thresholds and the MITs in particular. In chapter 7, the concept of the mean time between false alarms (MTFA) will be introduced and discussed in detail. In that section, it will be shown how the MTFA will render MITs perfectly suitable for acting as alarm thresholds in AAL settings.

6.4.   Advanced Alarming Criteria

6.4.1.   Preliminary Considerations about Advanced Criteria

In the previous section, alarm thresholds that only address overall inactivity in the flats were considered. While it is indeed feasible to raise alarms based solely upon the actual inactivity exceeding a maximum permissible inactivity threshold, more sophisticated alarm rules are conceivable and should be put in place to increase the overall selectivity of the monitoring and alarming system. Several additional techniques generating and providing additional knowledge about the state of the flat and the conditions of the tenant will be discussed in this chapter.

In case of a fall, it is anticipated that the inactivity pattern of the affected tenant will change – activity can either cease entirely if the tenant becomes unconscious or only motion detector signals will be captured after the fall if the tenant is still able to move or crawl. In the latter case, this pattern change might be detected and be utilised as an additional alarming criterion. However, user activity must be monitored rather closely in order to obtain sufficient information to allow timely reactions. This is contrary to the concept of gathering as little information about user behaviour as possible followed in the Kaiserslautern and Bexbach projects. In addition to pattern changes, the room in which the tenant resides may offer clues to possible emergency conditions: For example, a user being in the bathroom for several hours during day-time may hint at a potential emergency. However, in other rooms, e.g., the sitting room or the bedroom, thresholds cannot be as low as in the bathroom since legitimate sojourn times in those rooms can indeed be several hours, thus requiring alarm thresholds to be high enough to accommodate these regular use patterns. In the last section of this chapter, long-term pattern shifts will be assessed. It is expected that long-term patterns will bear traces of the onset and worsening of chronic illnesses or other phenomena occurring over long periods of time (e.g., changing user habits due to the season of the year). However, no chronic illness could be observed in the Kaiserslautern project so that hypotheses relating to chronic illnesses could not be verified.

6.4.2.   Pattern Changes in Case of a Fall

One pivotal question is what kind of activity pattern is observed in a flat after the tenant fell. In general, two distinct cases have to be considered: If the tenant faints or becomes unconscious, activity is likely to cease entirely. If, however, the tenant falls but does not rest motionlessly on the floor or if pets live inside the flat as well, there will still be activity being observed by the motion sensors. Especially in case of the user panicking it is to be assumed that he will try to draw attention to himself as quickly as possible, e.g., by trying to reach the phone or trying to crawl to the front door. The basic threshold approaches described above are well capable of covering the former case. In the latter case, however, they are likely to fail because either the user or their pets continue to trigger motion detectors and thus generate activity.

An approach to circumventing the problems connected to continued motion inside a flat after an emergency has occurred is interpreting singular events and continuous activity separately. When thinking about the two different types of activity data –continuous and singular– it can be assumed that in case of a fall or collapse, no intentional and wilful interaction with any device capturing singular events (e.g., wall switches, doors, or windows) can be performed by the tenant any more. In contrast to that, physical motion of the person lying on the floor or possibly present pets may still be recorded by the motion detectors. Hence, it may seem promising to introduce a second inactivity profile only taking into account singular events, thus possibly mitigating the issue of ongoing continuous activity: Regardless of the general inactivity level evaluated above, such a threshold will trigger alarms if no singular events are captured for a certain period of time, no matter whether or not motion had been observed by the motion detectors. It needs to be emphasised, however, that alarming thresholds based on singular events only cannot be used at night-time – during the sleep phase of the user, no singular events whatsoever are to be expected. Thus, the usability of such thresholds is confined to day-time.

Fig. 6.29 shall exemplify this approach. The scatter plot displays all data points recorded on a given day in flat A, differentiating five sources of captured data: motion detectors (i.e., *continuous* data) on the one hand, and windows, lights, roller blinds, and door (i.e., *singular* events) on the other hand.

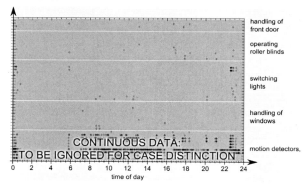

Fig. 6.29: Raw data captured in 24 hours; only singular events important for reasoning in this section

The total number of data points displayed in the above diagram amounts to approximately 2100. After having discarded all data points relating to continuous activity, approximately 70 data points remain, i.e., roughly three percent of the initial amount of activity information. If it is assumed that these remaining singular events were equally distributed over the day, on average every 1200 seconds, i.e., roughly every twenty minutes, a singular event should occur. In reality, however, the singular events are not ideally distributed, leading to phases of inactivity of significantly more than one hour (see Fig. 6.30).

Comparing Fig. 6.30 (a) and (b) corroborates the assumption that mainly the motion detectors contribute to the captured activity data in a flat. If the motion detector data are ignored as in diagram (b), the overall level of inactivity rises significantly from almost zero during the day to peaks of more than one hour with the longest inactivity peak lasting 2.75 hours.

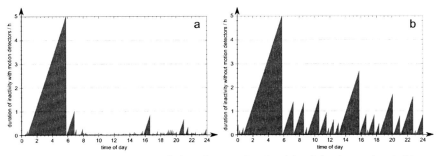

Fig. 6.30: Comparison of inactivity profile based on complete data (a) and singular events only (b)

Even though Fig. 6.30 only shows a single day's inactivity pattern and only moderate peaks of typically less than two hours, the omission of motion detector signals is by no means negligible. When establishing 28 day inactivity patters as introduced above, the below long-term singular inactivity pattern result (Fig. 6.31 and Fig. 6.32). All of them show several inactivity peaks of three hours and more, thus necessitating thresholds of whatever kind to be at least four hours. As explained in previous sections, thresholds with response times this long need to be ruled out for application in AAL environments.

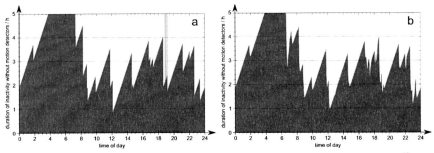

Fig. 6.31: Flat A: 28 day inactivity pattern considering singular events only; (a) maximum-based[16], (b) outliers removed and smoothed

In flat A (see above), even the 28 day pattern with outliers removed (b) exhibits numerous inactivity peaks of several hours. The highest peak in the considered 28 day period was exactly four hours. In case of flat B, the inactivity peaks rise even higher: Several peaks of more than five hours were observed (Fig. 6.32). In addition, numerous inactivity periods lasting between three and five hours occurred as well.

---

[16] The peak-shaped artefacts occurring at around 1900 in diagram (a) are spurious and caused by the retrospective data analysis performed in this work. They do not occur when processing the sensor signals in real-time and are thus ignored in this work.

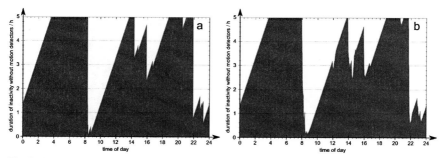

Fig. 6.32: Flat B: 28 day inactivity pattern considering singular events only; (a) maximum-based, (b) outliers removed and smoothed

The two possible ways of dealing with such long periods of inactivity are either lifting the alarm threshold to a response time long enough to make sure no false alarms occur or accepting a certain number of false alarms per unit of time: If whatever alarm threshold –be it based on an (m)SLAT or MIT– was shifted above the highest peaks observed during the initial 28 day training period, persons requiring help would have to wait for more than four and up to six hours for help to be administered. If, on the other hand, false alarms were to be tolerated and the threshold would be set to similar response times as in Fig. 6.25, several false alarms would be incurred. The following Table 6.4 gives the number of false alarms that would have been encountered with different mSLATs based upon singular events (subsequently referred to as *mSingSLATs*) being applied to the singular events' inactivity graphs shown in Fig. 6.31 and Fig. 6.32. In this table, the total number of false alarms is no more broken down to the individual causes (e.g., laying in) thereof because it cannot be determined when the sleep period of a person ends without utilising motion detector data.

Table 6.4: Number of false alarms in one month in flats A and B based on various mSingSLATs

flat A		flat B	
Threshold	**No. of false alarms**	**Threshold**	**No. of false alarms**
$mSingSLAT_{90/09\text{-}22}$	46	$mSingSLAT_{90/09\text{-}22}$	69
$mSingSLAT_{120/09\text{-}22}$	28	$mSingSLAT_{120/09\text{-}22}$	47
$mSingSLAT_{150/09\text{-}22}$	13	$mSingSLAT_{150/09\text{-}22}$	33
$mSingSLAT_{180/09\text{-}22}$	9	$mSingSLAT_{180/09\text{-}22}$	22
$mSingSLAT_{210/09\text{-}22}$	4	$mSingSLAT_{210/09\text{-}22}$	16
$mSingSLAT_{240/09\text{-}22}$	1	$mSingSLAT_{240/09\text{-}22}$	12
$mSingSLAT_{270/09\text{-}22}$	0	$mSingSLAT_{270/09\text{-}22}$	7
$mSingSLAT_{300/09\text{-}22}$	0	$mSingSLAT_{300/09\text{-}22}$	6
$mSingSLAT_{330/09\text{-}22}$	0	$mSingSLAT_{330/09\text{-}22}$	4
$mSingSLAT_{360/09\text{-}22}$	0	$mSingSLAT_{360/09\text{-}22}$	2

The above table underscores that it is very difficult if not impossible to solely utilise singular activity signals for identifying potential cases of emergency. The dilemma that cannot be easily tackled is balancing the response time and the number of false alarms. The above data suggest that no viable compromise can be reached. It has thus to be concluded that mSingSLATs are no effective means for enhancing the detection rate or shortening the response time of a given monitoring system. It needs to be noted, however, that this result is no universal truth – this finding is based on the actual implementation of the Kaiserslautern and Bexbach AAL project as described in subsequent chapters. If more sensors are available, as it is commonly the case in projects aiming at identifying activities of daily living (ADLs) [e.g., [Cook, 2006], the number of singular events should increase significantly, thus possibly rendering the mSingSLAT approach a valuable tool for detecting emergencies.

6.4.3. Occupied Room Criterion

The occupied room criterion (ORC) was devised to enable establishing additional alarm rules on top of merely monitoring inactivity levels. For the occupied room criterion to be operational, knowledge of the exact whereabouts of a person in their flat needs to be obtained. By analysing the whereabouts and its duration, inexplicably long sojourn times in certain rooms of the flat are to be detected that may be indicative of possible emergency situations the tenant could be in. For example, it is very unlikely that a person stays in the bathroom for several hours. It needs to be noted, however, that there are *limitations* to the ORC approach and that it can only be employed as an auxiliary monitoring technique accompanying inactivity analysis.

Some of the limitations of the ORC that need to be kept in mind are misleading sensor signals in case that there is a pet in the flat as well as certain technical properties of the sensors being used in the two pilot projects. In the former case, a pet might trigger sensors in other rooms of the flat than the one the user is actually in, thus inducing errors in the correct identification of the room inhabited by the user. This is an inherent, systematic problem that needs to be accounted for by the design of the AAL system. The latter problem is caused by the switch-off delay of the motion sensors used, i.e., the time between the actual cease of activity and the point in time at which the sensors signals this. Moreover, the effective detection range of the motion sensors may not always suffice to reliably detect a person if their distance from the sensor is too large. I.e., situations are conceivable in which a tenant is at home safe and sound but the motion detectors do not signal any activity, e.g., because the user sits on the sofa reading his newspaper. In such a case, a false alarm might be raised by the ORC when the user cannot be tracked reliably and the erroneously determined sojourn time for the assumed room exceeds the threshold at that respective time of the day. It is expected that future sensors specifically designed for the use in AAL settings will no more cause such problems. However, it is common to most sensors sampling a parameter or measurand such as motion that switch-off delays must be accommodated for. Thus, the FSMs and algorithms processing the sensor data must be able work on data that are collected with sensors with reasonable switch-off delays, e.g., up to 20 seconds. A switch-off delay of this duration is believed not to cause significant problems when processing and interpreting the data.

With regard to the aim of tracking the user's location, it is important to note that in the two pilot projects on which this work is based two different sensor technologies –KNX and EnOcean– with different technical characteristics are used. Both of them have certain pros and cons and both are thus tested in real environments to evaluate which technology is better suited for AAL applications in the long run. One of the technical side effects with substantial impact on location tracking is that both sensor technologies signal cease of motion or activity, respectively, only with a delay of 12 (KNX) or 1 to 100 (EnOcean) seconds, respectively. The following Fig. 6.33 illustrates the mode of operation of the two sensor technologies.

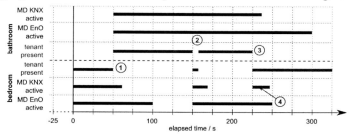

Fig. 6.33: Triggering and re-triggering characteristics and switch-off delays of KNX/EnOcean motion detectors (MD)

In the above example, let there initially, i.e., before 0 seconds at which this consideration begins, be no activity in both the bathroom and the bedroom. Let further at $t=0$ seconds activity caused by the tenant begin in the bedroom. Both a KNX and an EnOcean motion detector will signal activity. Let the tenant at $t=50$ seconds adjourn to the bathroom (marker 1). At this point, a concurrency situation will set in: Both types of motion detectors will signal start of activity in the bathroom. At the same time, the motion detectors in the bedroom will not yet have signalled stop of activity within their detection range. Even though the delayed stop signals are not ideal, they do not pose too big a problem if the tenant does not switch rooms several times within a short period of time. However, due to the two different signalling strategies implemented in the KNX and EnOcean motion detectors, the delay before stop of activity is signalled can differ considerably. While the KNX sensors signal inactivity 12 seconds after the last activity had been observed, the EnOcean sensors have a fixed, built-in sampling rate and scan their environment every 100 seconds. Hence, in the worst case it could take up to 99 seconds for them to notify the monitoring system of the end of activity if the tenant switched rooms just after the EnOcean sensor scanned its environment. Regarding the room change marked by (1), however, the whereabouts of the user will correctly and unambiguously be determined 50 seconds after the room change at the latest, i.e., when the EnOcean sensor in the bedroom will switch off. Marker 2 illustrates another instance of this effect: Let the tenant enter the bedroom at $t=150$ seconds only for a few seconds, e.g., to fetch something or turn off the light – neither a KNX nor an EnOcean sensor in the bathroom can detect the brief absence from the bathroom but both sensors in the bedroom will signal activity. After having returned to the bathroom, there will again be a fuzzy situation in which the location of the user will be unclear. When finally returning to the bedroom at t=225 seconds (marker 3), an EnOcean sensor will still be in its active state from the previous visit (marker 2). In contrast to that, a KNX sensor will have switched off in the meantime and will be triggered again. Let the tenant eventually sit down near the window outside the range of the motion detector for a prolonged period of time (marker 4). This will lead to all sensors switching off, both in the bedroom and the bathroom, no matter what technology they use. In this case, it will not be possible to be determined where the tenant is because he could as well have moved back to the bathroom and be out of reach of the motion detectors there.

Even though yet more complex situations are conceivable that cannot be discussed within the scope of this work, the above example clearly illustrates the difficulties connected to identifying the exact location of the user with a very high temporal resolution. In conclusion, it must be acknowledged that exactly identifying the user's location (i.e., room) at times when several room changes occur within a short time is hardly possible. Particularly the fact that the flat is not perfectly covered by the ranges of the motion detectors exacerbates the location tracking task.

However, these shortcomings are only of circumstantial relevance for real-world AAL applications. Since switching rooms indicates substantial activity –the tenant is well capable of moving around–, there is no reason to assume that any alerts need to be raised shortly after having observed any room changes. Thus, the fuzziness inherent to location tracking does not hamper emergency detection as it inevitably implies activity at times when it occurs which in turn most likely means that everything is alright.

At other times, when the user remains in a room not only for a few seconds or minutes but for several minutes or longer, location tracking does indeed work reliably and provides valuable information even though tracking the user closely and accurately when changing rooms is challenging. Due to the fact that all motion detectors switch off and re-trigger continuously (see raw data plot in Fig. 6.29), faulty reasoning caused by multiple room changes will automatically be corrected as soon as a user stays within the same room for a few minutes and re-triggers a sensor. It can thus be reliably deduced in which room the user remains for long periods of time.

Fig. 6.34 depicts the finite state machine[17] used for user tracking. For reasons of simplicity and comprehensibility, the graphical representation of the FSM has been simplified: First, in contrast to the presence FSM shown in Fig. 4.5 the active room FSM has no time constraints. Thus, the transition conditions do not contain any information about time constraints. Second, window operated can mean both *window opened* and *window closed*. In case that there is more than one window in a room, it refers to all windows in that room. Third, all transition conditions pointing to one of the three rooms can only be triggered by sensors located in the room they are pointing to. I.e., the transition condition motion detector fires | bedroom belonging to the transition from the sitting room/kitchen to the bedroom can only

---

[17] Please refer to section 4.2.4 for an explanation why Mealy automata were preferred over Moore automata.

triggered by the motion detector *in the bedroom.* Fourth, a new kind of event has been introduced. Events containing an "@" mean that they are triggered by output of other FSMs, i.e., `front door open @ Presence FSM` is a transition that can be triggered if the respective output is generated by the presence FSM, thereby linking multiple FSMs to each other. Finally, it needs to be noted that in case of the EnOcean sensors –which routinely transmit their current state at fixed intervals– a state transition in the automaton can only be invoked on the first occurrence of a telegram indicating a change of a physical sensor state, i.e., recurring routine telegrams cannot trigger a state transition. Lights and roller blinds are not considered as events for user tracking as they can either be triggered directly in the respective rooms by operating wall switches but also by remotely controlling the devices via a central control unit PAUL from other rooms. The captured telegram is, however, identical in both cases. Hence, these events do not allow reliable conclusions about the location of the user.

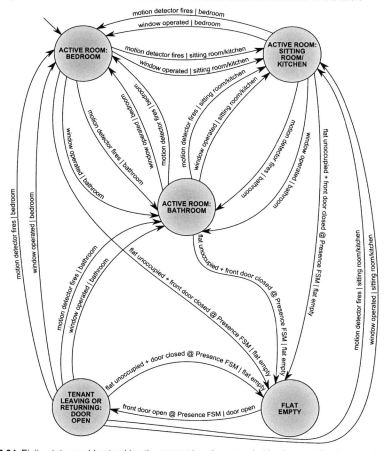

Fig. 6.34: Finite state machine tracking the current location occupied by the user ("active room")

The above active room FSM allows monitoring where the user is located at any given time provided that the user had not been moving from room to room several times in the last minutes. The task of this FSM is explicitly not detecting inactivity which is instead covered by MITs and (m)SLATs. Using the active room FSM, it is possible to keep track of how long a person resides in a specific room. This allows enhancing the inactivity monitoring approach described above and its shortcomings regarding the detection of emergencies in which the user does not lie still, e.g., after a fall. Assessing the additional knowledge provided by the active room FSM, this deficiency of the inactivity monitoring approach can partially be mitigated: It is to be assumed that a person who fell and is now lying on the floor may move or try to crawl but will not sojourn multiple times to other rooms. Thus, staying for inexplicably long times in the same room may hint at a fall or the like. Individual thresholds for different rooms are conceivable, e.g., it is less likely to spend two hours on end in the bathroom than in the sitting room/kitchen. Another example are nightly bathroom visits – if the tenant enters the bathroom at night, coming from the bedroom, and remains in the bathroom for more than a given time, a case of emergency is to be assumed. The following diagram (Fig. 6.35) shows sample data from flat A, indicating the periods of time the tenant spent in the different rooms of their flat as determined by the *active room FSM* and *not* durations of inactivity.

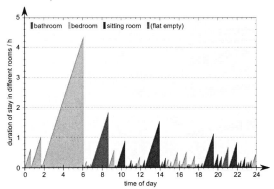

Fig. 6.35: Duration of stay in various rooms throughout the day (flat A)

The above diagram clearly shows that the tenant did not remain in a specific room for more than 1.5 hours before moving to another room during day-time on the particular day displayed. Moreover, there had been only a single period of time spent in the bathroom that was long enough even to be discernible in the above diagram – all other visits to the bathroom were so short that they cannot be made out in the above diagram. In Fig. 6.36 and Fig. 6.37, more thumbnails showing data from flats A and B are displayed to give a long-term overview of the typical sojourn times of the tenants in different rooms.

The two sets of diagrams below (Fig. 6.36 and Fig. 6.37) showing the duration the users stayed in the rooms of their flats give a good visual impression of what sojourn times are typically to be expected in flats A and B. On closer inspection of the data sets from the two flats, both similarities and dissimilarities stand out.

Regarding the similarities, it turned out that during day-time (0900–2200, in accordance with the PO of all inactivity alarm thresholds above) an abundance of room changes can be observed in both flats. As a result, in the vast majority of cases the sojourns in a specific room are of very short duration. Few sojourns, however, are of longer duration but typically less than two hours. When investigating the time spent in the individual rooms in more detail, it is unsurprisingly found that the amount of time spent in a room greatly depends on the type of room: In both flats, the bathroom proves to be the room the least time is spent in.

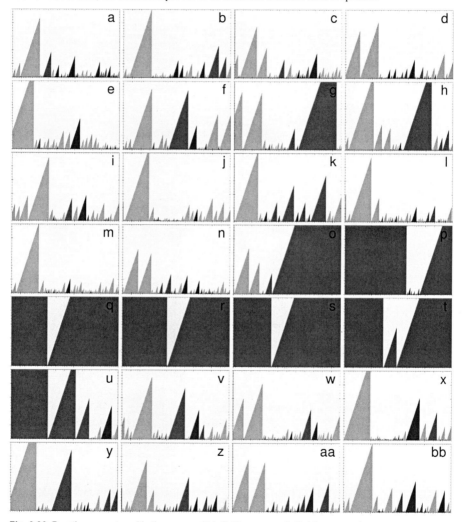

Fig. 6.36: Duration user stayed in the rooms of his flat in one month (flat A, summer)

When looking at the other rooms, however, significant differences between the users and their behaviour become apparent. The user in flat A spends a considerable amount of time in their bedroom during the day, whereas the user in flat B resides almost exclusively in the sitting room/kitchen (these two rooms are combined in the block of flats of the pilot project, similar to an open-plan apartment). This discrepancy can be attributed to the fact that user A has set up their working space in the bedroom so that the bedroom is not only being used for sleeping but also frequently for all kinds of activities throughout the day.

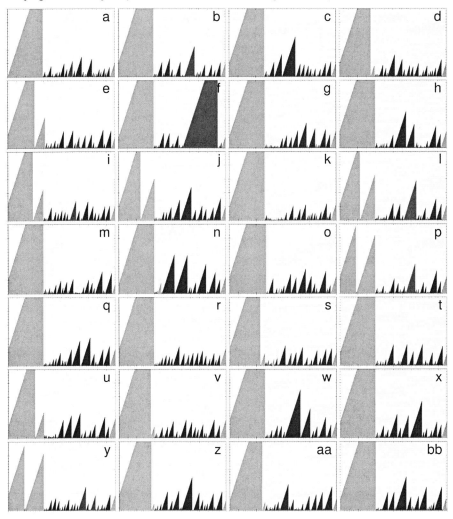

Fig. 6.37: Duration user stayed in the various rooms of his flat in one month (flat B, summer)

Such individual living conditions of the inhabitants of different flats need thus to be taken into account when designing AAL solutions to allow the representation of a wide range of different user behaviour and, as a result, make AAL technology as universally usable as possible.

The following two diagrams (Fig. 6.38) show a detailed breakdown of the maximum sojourn times of the users in flats A and B over a period of four weeks. This comparison supports the above conclusions that both similarities and dissimilarities can be found in the data from different flats. One feature both data sets have in common is that the time spent in the bathroom is very low. In addition, the time spent in the bedroom is also very low in flat B. In contrast to that, considerable periods of time are spent in the sitting room (flats A and B) and the bedroom (flat A only).

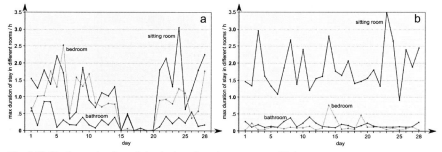

Fig. 6.38: Maximum duration of stay broken down to rooms over 28 days, period considered 0900-2200 (day-time): (a) flat A (summer), (b) flat B, (summer)

Based on the above maximum sojourn time graphs, new alarming criteria can be introduced which can –used in conjunction with the inactivity threshold used for raising alarms described in the previous section– increase the ability of the AAL monitoring system to detect possible emergency situations. The most discriminatively powerful criterion can be based on the time spent in the bathroom. Moreover, there is reason to assume that the use of the bathroom does not differ substantially between day-time and night-time. The second best criterion will evaluate the time spent in the bedroom – depending on the flat considered, this criterion is also assumed to be quite powerful (e.g., in flat B). The sitting room, however, does not allow establishing a very strict alarming criterion since people tend to spend significant amounts of time there every day, often several hours on end. Hence, the time spent in the sitting room can rather only be applied as a fallback criterion that becomes effective if the tenant should fall in the sitting room and was still able to move, thus preventing the inactivity thresholds from triggering an alarm.

On the basis of the above evidence, three ORCs for each of the flats A and B were defined.

Table 6.5: Occupied room criteria for raising alarms in different flats

Flat A		Flat B	
bathroom	$ORC_{BATH/00-24}$ = 60 min	bathroom	$ORC_{BATH/00-24}$ = 30 min
bedroom	$ORC_{BED/9-22}$ = 120 min	bedroom	$ORC_{BED/9-22}$ = 60 min
sitting room/ kitchen	$ORC_{SITTING/9-22}$ = 180 min	sitting room/ kitchen	$ORC_{SITTING/9-22}$ = 180 min

The actual alarm principle in which the ORCs are embedded is formally represented below. Other than the inactivity thresholds that are triggered if the overall inactivity exceeds a limit, the ORCs do *not* monitor inactivity but sojourn times in the rooms of the flat, i.e., the times that elapse before a room change occurs. An alarm will be raised if the sojourn time at any given time of the day exceeds the permissible time defined for the particular room the user is in.

Definition 6.4: Alarm principle of ORCs

IF   (*time of day t is outside PO*)
THEN   automatic alarming is disabled

ELSE   **IF**   (*sojourn time in room X exceeds $ORC_{X/PO}$*)
           THEN   a potential emergency is to be assumed **AND**
                      further steps need to be taken

In order to verify the fitness of the ORC alarm principle, the number of false alarms in flats A and B was determined using the data from the same months as in section 6.3. Table 6.6 lists the numbers of false alarms that would have resulted if the ORCs had been active in the respective months.

A string of conclusions can be drawn from Table 6.6. First, it is very important to note that several spurious telegrams had been received in both flats (4 in flat A, 2 in flat B). Especially the motion detectors in the bathroom fired for no apparent reason during night-time, thereby causing the active room FSM to switch to the state `active room: bathroom`. However, because the tenants were in fact fast asleep, no telegrams triggering a transition back to `active room: bedroom` could be received. Hence, the spurious telegrams and the erroneous transitions to the bathroom would ultimately have caused false alarms since the AAL monitoring system would have assumed that the users had remained in the bathrooms for hours but in fact never got out of their beds. Second, it turned out that the number of false alarm caused by the ORC differed significantly between the two flats: 19 false alarms in 4 months would have occurred in flat A in contrast to 19 false alarms in 10 months in flat B. This difference can be attributed to the observation that the daily routine of the tenant in flat B is much more uniform than that of the tenant in flat A. It can thus be concluded that having a fairly constant daily routine contributes to the usability and detection reliability of an AAL system. As soon as there are substantial deviations between the learned behavioural pattern and the actual single-day pattern observed, rule-based approaches will inevitably trip more false alarms.

Table 6.6: Number of false alarms in flats A and B based on ORCs for different rooms and POs

Flat A					Flat B			
	Total number of false alarms	$ORC_{BATH/0-24} =$ 60 min	$ORC_{BED/9-22} =$ 120 min	$ORC_{SITTING/9-22} =$ 180 min		Total number of false alarms	$ORC_{BATH/0-24} =$ 30 min	$ORC_{BED/9-22} =$ 60 min	$ORC_{SITTING/9-22} =$ 180 min
S1 08	3	1 (1: †)	2	0	S1 08	3	0	0	3
W1 08	4	1 (1: †)	3	0	W1 08	1	0	0	1 (1: †)
S1 09	10	3 (2: †)	6	1	W2 08	0	0	0	0
S2 09	2	0	1	1	W1 09	0	0	0	0
					W2 09	2	1 (1: ‡)	0	1
					S1 09	3	0	0	3 (2: †)
					S2 09	2	0	0	2
					W3 09	4	1 (1: †)	0	3
					W4 09	2	1	0	1
					W5 09	2	1 (1: †)	0	1

(†): Number of false alarms in parentheses caused by malfunctioning of sensors or spurious telegrams
(‡): Number of false alarms in parentheses caused by loss of telegram

Third, it must be acknowledged, however, that upon closer, differentiated inspection the three ORCs seem unequally fit for the desired purpose of being auxiliary alarming thresholds: For example, it can be considered a certainty that the bathroom ORC is universally suitable since it leads to very few false alarms (9 cumulative false alarms in 14 months, 7 of them caused by technical flaws of the used sensor equipment), can be active 24/7, and can be set to response times of one hour or less, whereas for instance the sitting room ORC cannot be applied as easily. The fitness of the latter greatly depends on the individual behavioural pattern of the tenant: In flat A, only two false alarms would have been tripped by the sitting room ORC whereas 15 false alarms would have occurred in flat B – this can be put down to the fact that tenant A spends most of his time in the bedroom whereas tenant B resides a lot more in the sitting room. Conversely, the bedroom ORC works very well for flat B but would cause a substantial number of false alarms in flat A.

Finally, it needs to be kept in mind that the false alarms raised by the ORCs will occur in addition to those triggered by the inactivity threshold put in place. In case of flat A, the total number of false alarms resulting from $MIT_{9\text{-}22}$ and three ORCs would have amounted to 19 + 8 = 27 false alarms in four months; in flat B, there would have been 19 + 3 = 22 false alarms in 10 months. This result corroborates the above conclusion that uniform daily routines are beneficial for AAL alarming schemes – not only were there fewer ORC alarms in flat B than in flat A but also was the number of false alarms triggered by the inactivity rule lower as well. In conclusion, it can be stated that ORCs do indeed have the potential of improving the ability of the AAL system of detecting potential cases of emergency. Especially the bathroom ORCs response time can be set reasonably low so that a substantial improvement over inactivity thresholds only can be achieved. The other two ORCs feature longer response times but are still worth being considered. By allowing the individual users to choose which ORC is to monitor their daily routine and what thresholds to apply, self-determination of the users can be ensured as well as the number of false alarms can be adjusted to the users' preferences.

6.4.4.   Long-term Pattern Shifts

6.4.4.1.   Basic Considerations

Up to now, only reference patterns were created based upon 28 days of sample data. This sample data was taken from the beginning of the data capturing period, i.e., from four weeks during which contiguous data was recorded after the AAL system had been installed in a flat. The reference patterns as well as the inactivity thresholds and active room criteria derived therefrom are based on 28 days due to the reasoning that 28 days constitute a reasonable teach-in period that suffices to observe most of the typical behaviour a tenant exhibits. Longer learning phases were refrained from because they would thwart the usefulness of the AAL system as the system is not operational, i.e., not being able to trigger alarms, during the learning phase.

In this section, one of the main questions to be answered by long-term data scrutiny is whether and how slow changes and trends might be exploited for detecting anomalous user behaviour potentially hinting at early stages of medical conditions. It is to assumed, however, that such slow pattern changes can have multiple causes. Reasons for such pattern changes include daylight saving time, changing seasons (which in turn lead to changing needs for artificial lighting), changing working shifts, or general changes in the daily routine without apparent reason, to name but a few. Such changes might in turn necessitate updating the reference patterns (SLATs, MITs, ORCs) used for alarming as they may no more represent the typical user behaviour.

In order to determine whether or not any significant slow, gradual changes in the users' behaviour occur over long periods of time, multi-day patterns for *all* months that were so far only used for validating the novel inactivity thresholds, were now examined. Since it turned out in the course of this data interpretation step that long-term trends are not readily identifiable, data from two more flats had to undergo scanning for long-term changes as well. In order to limit the number of diagrams to a reasonable number, only the data from flats A and B will be visualised in full detail. In case of the two additional flats, only the final bar charts will be displayed.

6.4.4.2.   Chronic Medical Conditions identifiable by Long-term Pattern Shifts

In a very general sense, *chronic illnesses* are often characterised by, e.g., complex causality, long latency periods, or prolonged course of illness. In the long run, they may lead to functional impairment or disability [AIHW, 2008].

Since this work is based on ambient sensors only, vital signs cannot be monitored. Hence, only medical conditions that are related in any way to the activity and inactivity patterns of the persons affected can be identified. Although this constraint limits the number of illnesses that may be detected, several examples of illnesses identifiable with AAL technology in one way or another are discussed in the technical literature. Amongst those are dementia, depression, chronic fatigue syndrome/occupational burnout, diabetes, arthritis, osteoporosis, and certain types of heart diseases or cancer (e.g., [AIHW, 2008, Ekstedt et al., 2006, Ohashi et al., 2004, Roscoe et al., 2007, Shaji et al., 2009]). It needs to be kept in mind, though, that there may be reciprocal effects, i.e., illnesses both causing and being caused by reduced physical inactivity. In other words, situations are conceivable in which observed changes in physical activity are not caused by an actual illness but may be causative factors for the manifestation of a future chronic illness. Table 6.7 lists details regarding the parameters that are indicative of the illnesses named above. Even though this list is not to be considered comprehensive, it shows the broad range of chronic illnesses that may be addressed with AAL technology and the potential for improving peoples' quality of life by giving them the freedom to live independently at home as long as possible.

Identifying those medical conditions, however, is only possible if the individual who may potentially suffer therefrom is monitored over extended periods of time and does in fact develop the respective illnesses. Down to the present day, none of the participants in the pilot projects developed any of the aforementioned medical conditions. Even though this may be considered detrimental for the progress and success of the research work, it must be pointed out that it is undoubtedly the greater good that the tenants are in good health even if not all hypotheses and data mining algorithms can be verified based on empirical data. Thus, is has to be acknowledged that the data captured in the pilot projects is not perfectly suited for mining for features connected to the onset or course of chronic illnesses.

Table 6.7: Sample selection of chronic illnesses potentially identifiable with AAL technology

Dementia	+ A study aiming at identifying behavioural and psychological symptoms of dementia –namely Alzheimer's– revealed that there is a high prevalence thereof in dementia patients. Amongst the behavioural changes that were observed are wandering (68% of the monitored individuals), purposeless activity (50%), diurnal rhythm changes (55%), or sleep disturbances (40%). Activity monitoring in AAL environments is believed to be a suitable means for detecting such behavioural changes [Shaji et al., 2009].
Depression	+ Physical inactivity is a determinant [AIHW, 2008]
	+ Depression is often related to dementia. Thus, there is reason to assume that detection algorithms overlap [Shaji et al., 2009].
Chronic fatigue syndrome (CFS)/ Occupational burnout	+ Individuals suffering from CFS and subjects from a healthy control group show significant differences in their diurnal activity patterns. Those having CFS are likely to interrupt physical activity abruptly, possibly due to substantial fatigue [Ohashi et al., 2004].
	+ In patients suffering from occupational stress and burnout syndrome, sleep fragmentation and arousals, lower sleep efficiency, and thus pronounced sleepiness at day-time were observed [Ekstedt et al., 2006].
Diabetes	+ Physical inactivity is a determinant for Type 2 diabetes [AIHW, 2008]
Arthritis/ Osteoporosis	+ Physical inactivity is a determinant [AIHW, 2008]
Cancer	+ Amongst the typical symptoms of cancer are sleep disorders, e.g., problems falling asleep or maintaining sleep, limited regeneration, early wake-ups, or significant day-time sleepiness and restlessness. Potential causes for those ailments are biochemical processes connected to anti-cancer treatment and other symptoms frequently connected to cancer, e.g., pain, fatigue, or depression [Roscoe et al., 2007]. Although some of these symptoms will only occur after having diagnosed cancer and treatment has been initiated, they also seem to be prevalent in patients who are not currently being treated, e.g. cancer survivors. In general, restlessness and day-time sleepiness are believed to be easily detectable using AAL equipment if potential patients can be monitored closely, i.e., if sufficient sensor data is available.
Heart diseases	+ Physical inactivity is a determinant for ischaemic heart disease [AIHW, 2008]

6.4.4.3.   Long-Term Pattern Shifts based on Complete Sensor Data

In order to assess whether any long-term trends –even if they are not directly connected to chronic medical conditions– can be identified, the complete sensor data (as opposed to singular events only) from four flats (A, B, D, and E) will be scrutinised in this section.

Fig. 6.39 shows the graphs representing the maximum duration of inactivity based on the data collected in flat A whereas Fig. 6.41 illustrates the respective diagrams from flat B.

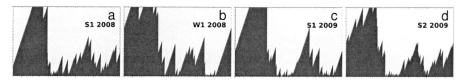

Fig. 6.39: One-month maximum-based long-term patterns from flat A, based on full data

In Fig. 6.39, no significant shifts in the one-month maximum inactivity patterns can be observed. The overall pattern exhibits a notable constancy: Long periods of inactivity from 0000 to approximately 0800, low inactivity in the morning from 0800 to 1300, inactivity periods of up to three hours between 1300 and 1800, reduced inactivity from 1800 to 2100, and increasing inactivity after 2100. There are, however, few exceptions to this classification. In diagram (b), there was one inactivity peak of more than two hours at 1130. Moreover, the flat sections of the graph between 0800 and 1300 as well as 1800 and 2100 that are visible in (c) are not always clear and pronounced. In diagram (c), they are very distinct, whereas only one of the flat sections occurs in (b) and (d). In (a), these flat sections are hardly discernible. In general, however, the durations of inactivity encountered during day-time throughout the four months are much alike. In conclusion, it needs to be stated that the four diagrams exhibit significant similarities so that obvious long-term shifts cannot be identified.

Table 6.8: Three longest periods of inactivity per month in summer and winter, flat A, full data[18]

summer, *PO* = [0900, 2200]				winter, *PO* = [0900, 2200]			
month	longest inactivity peaks / min		avg / min	month	longest inactivity peaks / min		avg / min
S1 2008 (a)	172	159	106	**146**	W1 2008 (b)	171	143	132	**149**
S1 2009 (c)	214	157	129	**167**				
S2 2009 (d)	150	149	80	**126**				

To substantiate the above visual conclusions, the three longest inactivity peaks during day-time (0900-2200) in these months were determined and compared. Since it was generally assumed that differences between summer (April to September) and winter (October to March) may be identified due to changed lighting and ventilation habits, Table 6.8 and Fig. 6.40 show the results separately for summer and winter to facilitate identifying season-related user behaviour.

Fig. 6.40 supports the above finding that no significant long-term changes having an impact on long inactivity periods occurred in flat A – the average duration of the three longest inactivity peaks encountered in the different months do not suggest any trends. In summer, the overall mean value of the three months from which data are available is almost identical to the value available in winter. It needs to be noted that in the following *grey areas* behind the bars in the diagrams indicate the *2σ-bands* of the overall mean value of summer and winter season.

---

[18] The "avg" column in this table and the following tables of this kind represents the average value of the three longest inactivity peaks, not that of all inactivity in the respective months.

Assuming that the duration of the inactivity peaks is normally distributed, the probability of a random value differing more than 2σ from the expected mean value is less than 5%. Thus, differences between winter and summer values are only considered statistically significant if they do not fall in each others 2σ-band.

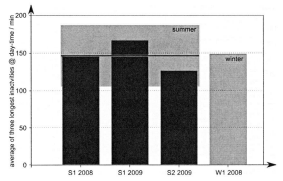

Fig. 6.40: Avg. of three longest inactivity periods (flat A); mean of summer/winter w/ 2σ range, full data

In order to further investigate this matter, more comprehensive data sets from flat B (Fig. 6.41) were examined in the same way. Overall, similar observations to those in flat A regarding long-term changes of the inactivity patterns were made in flat B (Fig. 6.41) as well. As diagrams (a) to (j) show, a constant low inactivity profile was captured in the flat during day-time (0900-2200). Apart from very few exceptions, all inactivity peaks during day-time last less than 90 minutes.

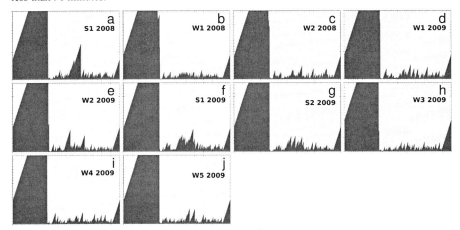

Fig. 6.41: One-month maximum-based long-term patterns from flat B, based on full data[19]

---

[19]Four days were discarded for the generation of diagram (f) due to a failure of the sensor array, thus having lead to faulty data capturing.

There is, however, marginal evidence that the durations of inactivity are slightly longer in summer than in winter. As in case of flat A, the three longest inactivity peaks are listed in the below table.

Table 6.9: Three longest periods of inactivity per month in summer and winter, flat B, full data

summer, *PO* = [0900, 2200]				winter, *PO* = [0900, 2200]			
month	longest inactivity peaks / min		avg / min	month	longest inactivity peaks / min		avg / min
S1 2008 (a)	155	83	83	107	W1 2008 (b)	41	33	33	36
S1 2009 (f)	97	76	71	81	W2 2008 (c)	59	48	47	51
S2 2009 (g)	67	67	62	65	W1 2009 (d)	64	53	46	54
					W2 2009 (e)	95	72	43	70
					W3 2009 (h)	40	40	39	40
					W4 2009 (i)	52	51	45	49
					W5 2009 (j)	66	66	43	58

Other than in flat A, the inactivity peaks (Table 6.9) do suggest that there is a marginal difference in the maximum inactivity durations between summer and winter. When looking at the two other flats D and E (Fig. 6.43 and Fig. 6.44) considered for this long-term data screening, however, this trend cannot be corroborated.

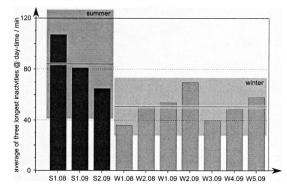

Fig. 6.42: Avg. of three longest inactivity periods (flat B); mean of summer/winter w/ 2σ range, full data

In the below diagram showing data from flat D, no significant difference between summer and winter months can be found. The observed user behaviour in principle corresponds to the one encountered in flat A, notwithstanding the notable difference in the absolute duration of the inactivity peaks (approx. 150 mins in flat A as opposed to approx. 80 mins in flat D).

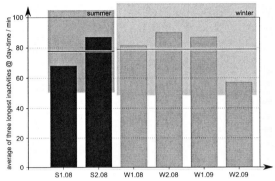

Fig. 6.43: Avg. of 3 longest inactivity periods (flat D); mean of summer/winter w/ 2σ range, full data[20]

Finally, the data from flat E exhibits yet another trend, converse to the one observed in flat B: In flat E, the maximum durations of inactivity encountered in summer are considerable shorter than those observed in winter. In S2 2009, the maximum duration of inactivity amounted to only half of those encountered in winter months. In S1 2009, the inactivity level was also lower than in winter. It needs to be noted that –other than in flats A, B, and D– the period of operation in flat E was chosen as $PO = [1000, 2100]$ due to significantly differing sleep/wake cycles.

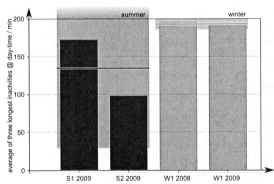

Fig. 6.44: Avg. of 3 longest inactivity periods (flat E); mean of summer/winter w/ 2σ range, full data[21]

The above examples underscore that simply looking at the maximum duration of inactivity over extended periods of time does not appear to constitute a reliable measure for identifying long-term pattern changes. Three different observations were made when examining the data from four flats, two of them standing in stark contrast with each other regarding winter/ summer inactivity durations: In two cases, no significant differences between summer and winter were observed, whereas in one case each the peak durations of inactivity were higher and lower, respectively, in summer than in winter. Thus, the overall inactivity level does not ex-

---

[20] Two days each were discarded in W1 2008 and W2 2009 (see footnote 19).
[21] Two days were discarded in W1 2008 (see footnote 19).

hibit generally exploitable features for classifying long-term behaviour of the users. In the following sections, the contributions of the various sensor types (light switches, roller blinds, doors, windows) to the overall data pool will be investigated in more detail in order to determine whether individual sensors may help to classify changing user habits.

6.4.4.4.   Breakdown of the Contribution of the individual Sensor Types to the Overall Sensor Data Mix

In the previous section, the dependence of the maximum inactivity durations based on the complete sensor data in one month on the season of the year had been investigated. There was, however, no clearly identifiable correlation of inactivity level and time of the year.

One may be inclined to believe that there are two opposing effects that might be called to explain the above results. First, it seems reasonable to assume that in winter, light switches are used much more frequently than in summer, thus increasing the overall inactivity level, i.e., decreasing the maximum durations of inactivity. Second, a possible counter-effect could be that in summer windows are opened more often to ventilate the flat or that roller blinds are operated to keep out the sun.

The following four diagrams show the average daily tripping rates of all of the various types of sensors for the same months that were considered in the previous section.

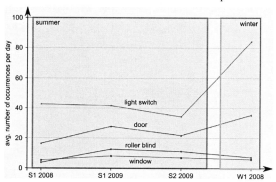

Fig. 6.45: Average daily tripping rates of different sensors in flat A

In Fig. 6.45, the daily tripping rates of the four sensor types *light switch*, *door*, *windows*, and *roller blinds* are displayed for four months. The tripping rates for the door, the windows, and the roller blinds are constant whereas the number of light switch actuations increased approximately by the factor 2 from summer to winter.

Fig. 6.46 shows similar trends. Again, the use of the door, windows, and roller blinds is fairly constant throughout the year. The increase in the actuations of the lights is even more pronounced than in flat A.

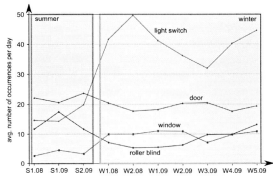

Fig. 6.46: Average daily tripping rates of different sensors in flat B

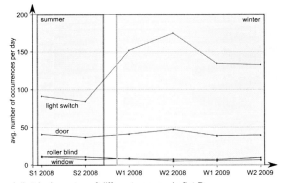

Fig. 6.47: Average daily tripping rates of different sensors in flat D

Fig. 6.47 also substantiates the observations made in flats A and B. Apart from the lights, all other sensors fire at a very constant rate throughout the year. In contrast to that, the tripping rate of the light switches is –as in the above cases– about 50% higher in winter than in summer.

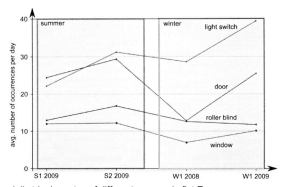

Fig. 6.48: Average daily tripping rates of different sensors in flat E

Fig. 6.48 is slightly different from the other three tripping rates diagrams above. The increase in light switch operations from summer to winter is not as pronounced as in the other cases whereas a notable drop in the door utilisation occurred in W1 2008. The general trend found so far is, however, also exhibited by flat E.

One of the **key findings** of the above data analysis is the observation that **both the absolute tripping rates and the peak inactivity times vary greatly between the flats but exhibit good constancy even over prolonged periods of time within the specific flat they were captured in.**

In conclusion, one of the two hypotheses introduced in the first paragraph of this section is corroborated by the detailed analysis of the individual activity data sources while the other is not. This means that the use of windows and roller blinds does not significantly differ in summer and winter, respectively. The number of light switch operations, however, does notably increase in winter, hence substantiating the initial assumption that lights are more frequently used when the number of sunshine hours is low. In Fig. 6.49, the peak inactivities are shown together with a *purely qualitative* graph (solid line) representing the utilisation of light switches in the respective flats to investigate whether or not there is any correlation between light utilisation and peak inactivity duration.

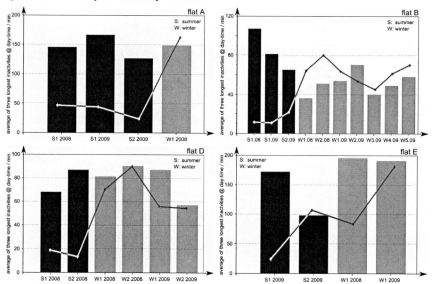

Fig. 6.49: Maximum duration of inactivity peaks and light utilisation (solid line) in flats A, B, D, and E

The above diagrams only indicate a limited correlation of the use of lights and the peak inactivity times encountered in a flat. In flat A, the inactivity durations are quite constant but light utilisation exhibits notable differences. The same holds true for flat D. In flat E, not only are the inactivity peaks of different duration but also does light utilisation change in an entirely incoherent way. The only flat in which a correlation of a nature whatsoever can be seen is flat B. In summer –when light utilisation is low– long inactivity peaks occur whereas in winter, when light utilisation is high, lower inactivity peaks were recorded. Due to the generally weak correlation, however, regarding any **conclusion made in a particular flat valid in other flats** does **not seem justified**. However, **within a particular flat**, certain trends can be observed, e.g., regarding pattern changes from summer to winter or vice versa. These trends may form the foundation for future long-term pattern analyses.

6.4.4.5.   Identified Long-Term Pattern Shifts based on Singular Events only

In this section, it is to be determined whether any long-term changes of inactivity graphs based on singular events only can be identified. As in the previous section, only data from flats A and B are visualised in full detail.

Fig. 6.50 illustrates the maximum duration of inactivity in various months considering singular events only. The overall increase of inactivity durations throughout the day compared to the diagrams in Fig. 6.39, comprising motion detector data as well, is clearly perceivable.

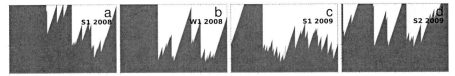

Fig. 6.50: One-month maximum-based long-term patterns from flat A, based on singular events only[22]

In line with the above way of representing peak inactivity, Table 6.10 lists the three longest periods of inactivity encountered in each of the months displayed in the above figure. As the diagrams already suggested, the duration of the inactivity peaks has increased substantially. The longest peaks now have a duration of up to six hours. Even the mean values of the three longest inactivities in a single month can now amount to more than five hours.

Table 6.10: Three longest inactivities per month in summer and winter, flat A, singular data only

summer, *PO* = [0900, 2200]				winter, *PO* = [0900, 2200]				
month	longest inactivity peaks / min		avg / min	month	longest inactivity peaks / min			avg / min
S1 2008 (a)	355	309	308	**324**	W1 2008 (b)	288	230	180	**233**
S1 2009 (c)	215	205	193	**204**					
S2 2009 (d)	360	282	281	**308**					

Fig. 6.51 graphically represents the values from the above table. Even though the average inactivity duration in summer is slightly larger than in winter, this difference is not statistically significant, both because the standard deviation (the grey area represents ±2σ) is very large and the population size is too small for drawing valid conclusions in a statistical sense. Based on Fig. 6.45, it would have been expected that the peak inactivity durations in winter

---

[22] Two days were discarded in diagram (a) (see footnote 19).

are considerably lower than in summer because of light switch usage that more than doubled in winter. This hypothesis could, however, not be verified by inactivity duration assessment.

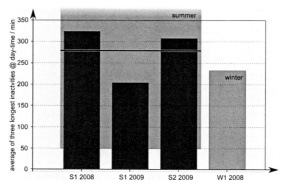

Fig. 6.51: Avg. of three longest inactivity periods (flat A); mean of summer/winter w/ 2σ range, singular events only

Fig. 6.52 shows the maximum inactivity graphs based on singular events only from flat B. Compared to Fig. 6.41, the increase of the maximum inactivity durations throughout the day is much more pronounced and significant than in case of flat A: In flat A, the peak inactivities doubled when omitting motion detector data whereas they quadrupled in flat B.

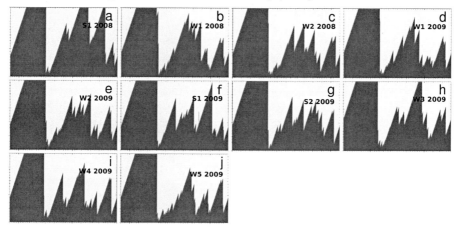

Fig. 6.52: One-month maximum-based long-term patterns from flat B, based on singular events only[23]

Table 6.11 gives a detailed breakdown of the inactivity peaks recorded in flat B when omitting motion detectors. In summer, the inactivity peaks amount to an average of 284 min-

---

[23] Two / four days were discarded in diagrams (b) and (f), respectively (see footnote 19).

utes and in winter, they amount to an average of 240 minutes. Fig. 6.53 graphically illustrates
the distribution of the inactivity peaks of the ten months listed in the below table.

Table 6.11: Three longest inactivities per month in summer and winter, flat B, singular data only

summer, PO = [0900, 2200]				winter, PO = [0900, 2200]				
month	longest inactivity peaks / min		avg / min	month	longest inactivity peaks / min			avg / min
S1 2008 (a)	404	395	331	377	W1 2008 (b)	277	244	244	255
S1 2009 (f)	293	236	224	251	W2 2008 (c)	232	231	227	230
S2 2009 (g)	233	218	217	223	W1 2009 (d)	264	225	219	236
					W2 2009 (e)	243	232	213	229
					W3 2009 (h)	347	259	223	276
					W4 2009 (i)	259	231	215	235
					W5 2009 (j)	239	227	206	224

As in case of flat A, the average duration of the inactivity peaks in summer is slightly larger than in winter, but since S1 2008 is likely to be an outlier, it has to be assumed that the
inactivity peaks in summer in winter are quite similar in terms of statistical significance.

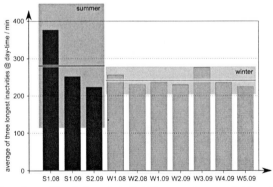

Fig. 6.53: Avg. of three longest inactivity periods (flat B); mean of summer/winter w/ 2σ range,
singular events only

The same holds true for flat D (Fig. 6.54). The data from the two summer months is rather
disparate, thus yielding a large standard deviation. As in the above two cases of flats A and B,
the average inactivity peaks' duration in summer is slightly larger than in winter. However,
even though this difference can still not be considered statistically significant due to the large
standard deviations, the data from three flats exhibit this trend.

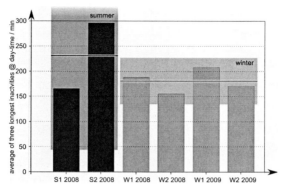

Fig. 6.54: Avg. of three longest inactivity periods (flat D); mean of summer/winter w/ 2σ range, singular events only[24]

When scrutinising the data from flat E (Fig. 6.55), an opposite trend becomes apparent: In summer, the inactivity peaks are shorter than in winter. Moreover, flat E is the only flat in which the difference in the peak inactivity durations between summer and winter may be considered statistically significant. This is clearly contrary to the findings from the other flats in which the inactivity peaks in winter tend to be shorter than in summer. However, the observed inactivity peaks including motion detector data and the ones for which motion detector signals were discarded match very well (compare Fig. 6.44 to the below diagram).

In addition, Fig. 6.48 shows that the distribution of singular events is more irregular in this flat E than in any of the other three flats, hence further exacerbating any attempts to find regularities in the inactivity data.

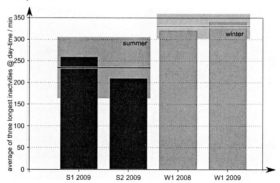

Fig. 6.55: Avg. of three longest inactivity periods (flat E); mean of summer/winter w/ 2σ range, singular events only[25]

Finally, one more **important observation** is to be pointed out: When comparing the two inactivity peak diagrams representing the long-term user behaviour in a flat –i.e., the one in-

---

[24] Two days each were discarded in W1 2008 and W2 2009 (see footnote 19).
[25] Two days were discarded in W1 2008 (see footnote 19).

cluding motion detector data and the one without this data–, it turns out that **in all flats those diagrams exhibit** at least some **concordance**. In case of flats B and D, the concordance can be rated fair, and in case of flat A, it is good and in flat E very good. Hence, it is to be concluded that omitting motion detector data does not generate a significant amount of new knowledge and thus that interpreting the complete sensor data set and only singular events separately does not constitute a straightforward way of further assessing the health status of the user.

Recapitulating the above results, several conclusions can be drawn from the long-term pattern analysis which are summarised below:

+ **Long-term patterns are unique to a particular flat. They represent the typical behaviour of the tenant in that flat. Long-term patterns generated in one flat cannot be transferred to another one.**

+ **In case of the flats evaluated in this work, the long-term patterns within a particular flat turned out to be fairly constant. However, long-term patterns of a user may change over even longer periods of time than those in this work so that alarm criteria based thereupon may need to be updated.**

+ **Although there may well be a causal relationship between certain features identifiable in the collected sensor data on the one hand and chronic illnesses of the tenants on the other hand, automatically drawing valid conclusions from the data about the actual cause of, e.g., an illness is not feasible. This is because a plenitude of causes may result in the same pattern or pattern change, i.e., in most cases pattern changes are not specific to a genuine cause.**

+ **Considering the above item 3, it must finally be noted that generic alarming based on long-term patterns does not seem to be a viable means for alarm generation. Evaluating and interpreting long-term activity and inactivity patterns always requires sound knowledge of the personal living conditions of the respective users.**

+ **However, even though the data patterns may not be indicative of a particular medical condition in themselves, they are expected to allow supporting or refuting hypotheses made by health professionals regarding the health state of the tenant. I.e., if a physician suspects that a particular medical condition may be present, the inactivity or activity patterns may help him to substantiate that assumption.**

It needs to be noted, however, that no chronic illnesses became manifest in the tenants having been monitored. Since the main focus of this work was establishing alarm rules for detecting emergencies rather than the gradual deterioration of the health status of a person, long-term pattern shifts will be addressed in further research to be conducted in the future. In order to analyse long-term behaviour in detail and be able to identify gradual changes, considerably longer streams of data are required as none of the individuals in flats A–E showed signs of a deteriorating health status during the period of monitoring. Furthermore, it must be ensured that the data is much more consistent, i.e., data capturing should not be interrupted ad libitum by the user even if the guidelines concerning the concept of complete self-determination had to be revised in that case (see section *8.5: Shortcomings* for more details on this matter). In addition, using more sensors for gathering additional information about singular activities and thus monitoring user behaviour more closely might help to identify long-term pattern drifts – several hours in between two singular events are too long a time to allow detailed analysis of the inactivity patterns of a user.

6.5.   Summary and Discussion

In this chapter, inactivity based alarm generation principles were introduced and compared to each other. First, the necessity for multi-day inactivity patterns was elucidated. Three different methodologies for computing such multi-day patterns were applied to the data captured in the pilot-projects: plain maximum-based, mean value-based, and outlier-free max-based 28-day long-term inactivity patterns. Twenty-eight days were chosen as the duration for computing the reference patterns since it is considered reasonable and feasible to train the AAL system for four weeks after initial deployment in order for it to learn the typical user behaviour. Longer training periods do not seem appropriate since they would mean that the system is not operational for prolonged periods of time which might not be acceptable for the users.

Upon closer examination, neither the plain maximum-based nor the mean value-based long-term patterns proved to be suitable for representing the long-term behaviour of the users or for forming a basis for inactivity monitoring and raising alarms based thereupon. On the one hand, the maximum duration of inactivity $MDI_p(\cdot)$ is too easily influenced by exceptional occurrences taking place during the training period, i.e., entirely unusual behavioural patterns, particularly extended inactivity peaks that are caused by typical user behaviour but owed to exceptional circumstances. On the other hand, the mean value-based $MDI_p(\cdot)$ smoothes the average user behaviour by far too strongly. Occasionally occurring inactivity peaks, e.g., an afternoon nap, are almost entirely removed from the long-term inactivity pattern. Removing obvious outliers by computing the four quartiles of the inactivity peak distribution and discarding all values greater than the upper quartile plus three times the inter-quartile range IQR as they are likely to be outliers (this procedure is typically used for creating *box plots*) helps to mitigate these problems. The resulting outlier-free inactivity reference pattern is eventually *slightly* smoothed, yielding $DIE_p^*(\cdot)$. $DIE_p^*(\cdot)$ is considered an appropriate representation of the long-term behavioural patterns of the users since it includes most inactivity peaks that occur during the training phase but drops inexplicably long ones.

Based on $DIE_p^*(\cdot)$, several alarming criteria, i.e., thresholds, were introduced. The most simple threshold, the static linear alarm threshold SLAT, was introduced as a benchmark. A SLAT constitutes a fixed threshold, active 24/7, based upon which an alarm is triggered if the current inactivity exceeds the SLAT. Due to the nocturnal long inactivity phases, the SLAT would have to be very high in order not to trigger an overabundance of false alarms. Thus, it is not reasonable to apply a SLAT as an alarming criterion since the response times would be unacceptably long. SLATs were thus further refined to mSLATs, i.e., masked SLATs. mSLATs are still linear thresholds, but are active only at certain times of the day (period of operation PO). With mSLATs, false alarms could be reduced to a reasonable number that are likely to be accepted by the user. mSLATs are thus considered suitable alarm criteria if the alarm system is not required to be operational 24/7. Finally, multi-day inactivity-derived thresholds (MITs) with a PO only covering day-time were introduced that take into account changing inactivity levels of a user throughout the day. MITs are derived from the outlier-free multi-day inactivity patterns are thus unique to the flat they were established in. The results from the Kaiserslautern project suggest that MITs are even better suited for alarm generation as they have two advantages over mSLATs: On the one hand, MITs can reduce the average response time considerably so that help can be administered more quickly in case of a genuine emergency. On the other hand, MITs can help to reduce the number of false alarms raised in a flat if too many occur per unit of time. In order to investigate whether MITs are suitable for 24/7 emergency monitoring, they were extended to a PO of 24 hours. Since the thresholds are established in such a way that no false alarms are triggered during the training phase, this led to unacceptably long response times, even though MITs are promising candidates for continuous monitoring. This issue will be addressed in the following chapter introducing the concept of the *mean time between false alarms* which means that a certain number of false alarms will be tolerated –already when establishing the MIT– for several reasons that will be explained below.

To further refine the alarming principles, advanced alarming criteria were researched as well in this chapter. It was hypothesised that in case of a fall or the like, a person might still be able to move or crawl but does indeed require help. Since the physical activity of the person needing help would trigger the motion detectors, simple inactivity thresholds might be ineffective as movement of the person would reset the inactivity peaks repeatedly. However, interpreting continuous activity (data from motion detectors) and singular events (all other types of sensors) separately did only partly mitigate this issue. Basing inactivity alarms on singular events only, i.e., triggering alarms if no singular event had been captured for a certain period of time, would require the thresholds to be very high, e.g., at the very least three hours or –depending on the user– much more. Response times of this duration render the monitoring system less useful since three hours should be the maximum response time in order to be able to administer help within a reasonable time. The large number of false alarms connected to this alarming criterion further exacerbates the situation so that applying this alarming criterion should be refrained from at this stage. Monitoring the user more closely in the future with a denser sensor array might help to mitigate the challenges connected to separating continuous activity and singular events.

The suitability of sojourn times in different rooms as an additional alarm criterion was assessed as well. For this reason, it was determined what the maximum sojourn times of a user in the different rooms of their flat were. It could be shown that especially the bathroom is a room in which the users do not spend more than 30 to 60 minutes on end. Thus, a feasible alarm rule is to raise an alarm if the user stays in the bathroom for more than one hour. The typical sojourn times in the other rooms vastly depend on the individual user but hardly exceed three hours and are typically less. Thus, the occupied room criterion ORC is well suited as an auxiliary alarm criterion, contributing to the discriminative power and detection rate of an AAL system.

Moreover, long-term pattern shifts are also of interest for AAL solutions. Knowledge about long-term changes in the behavioural patterns can be useful for two reasons. First, they may help to support or refute hypotheses regarding chronic illnesses and their course. Second, alarming criteria based on long-term patterns might have to be updated if the underlying inactivity profiles changed over time. Based on the data evaluation conducted within the scope of this work, however, none of the two possibilities could be observed. Features in the sensor data hinting at potential chronic illnesses were not found because no specific chronic medical conditions were prevalent in the individuals having been monitored. Moreover, the determination of the number of false alarms triggered by the (m)SLATs and MITs initially derived from four weeks' data collected after the deployment of the AAL monitoring system did not suggest that there was a significant drift in the activity/inactivity patterns exhibited by the users that warranted continuous updates of the alarm thresholds. The number of false alarms generated by the initial alarming criteria remained at low levels throughout the entire period considered in this work (4–10 months). Thus, inactivity thresholds of the above types are to be considered robust means for raising alarms in case of abrupt cease of activity in a flat. Another important factor to be considered regarding automatically updating any alarm criteria is that this concept potentially violates the requirement of self-determination of the users that must be adhered to at all times. If automatic threshold updates are not entirely transparent to the user, they can be the cause of unanticipated confusion and lack of acceptance.

To sum up, the below Table 6.12 lists the pros and cons of all the aforementioned alarming criteria from the author's point of view in a nutshell. *Ease of comprehension* means in this context that the users can easily understand the functional principle of the alarming criterion. *Representation of individual user behaviour* refers to the degree to which the alarm criterion adapts to user-specific peculiarities. The two columns referring to *robustness* indicate whether it is likely that occasional missing or spurious telegrams cause the system to malfunction or may lead to false alarms not triggered by actual user behaviour. Last but not least, *stochastic end of* activities indicates whether an alarming criterion is particularly susceptible to the fact that the point in time at which some of the EnOcean telegrams are sent cannot be predicted or anticipated exactly. Instead, those telegrams are sent at random times within an interval of 100 seconds (see section 6.4.3 for details on this matter). The actual ratings for the above properties range from "– –" for *very poor* to "O" for *fair* and "++" for *very good*.

Table 6.12: Comparison of the alarming criteria introduced in chapter 6

	Period of operation	Ease of comprehension	Representation of individual user behaviour	Robustness to telegram losses	Robustness to spurious telegrams	Stochastic end of activities	Outcomes in a nutshell
SLAT	24 h	++	– –	++	++	++	Possibly suitable for persons with overall very low inactivity. Otherwise very rough alarming criterion with either long response times or proneness to large numbers of false alarms.
mSLAT	Any	+	O	++	++	++	Same as SLAT, but because of the ability to restrict alarming to a limited period of operation, large numbers of false alarms at night-time due to the user being asleep can be avoided. Better adaptation to user-specific behaviour during day-time than SLATs.
MIT	Any	O	++	++	++	++	MITs best represent the users' inactivity profile throughout the day, i.e., inactivity peaks can be accommodated for by the MIT. Lower response times than (m)SLATs and/or fewer false alarms per unit of time.
mSing SLAT	Any	–	O	+	++	++	In general, singular events do provide information about the inactivity in a flat. However, singular events are far fewer than motion detector signals. Thus, response times turned out to be significantly longer than in case of the (m)SLATs. Can be auxiliary alarm threshold.
Occupied Room Criterion	Any	+	+	O (false alarms)	–	– – (faulty tracking)	Useful extension of the alarming scheme. Particular the bathroom proved to be worth being monitored with thresholds of 30–60 minutes.
Long-Term pattern changes/ complete data	n/a	+	O	++	++	++	Long-term patterns (LTPs) shall help in diagnosing and handling chronic illnesses. However, LTPs are not believed to allow direct conclusions as to which illness caused a particular pattern change. Instead, LTPs are meant to support or refute hypotheses made by medical professionals about the health state of a user. Fine-grained intra-day details of the user behaviour can only partially be covered by long-term pattern changes.
Long-Term pattern changes/ singular events only	n/a	–	O	++	++	++	

Finally, it should be noted that in order to demonstrate the full potential of the detailed analysis of comprehensive sensor data sets, so far only data from the Kaiserslautern project were used. The conceptual design of the Bexbach project aims at reducing the number of sensors to a minimum to minimise expenses for hardware and the installation thereof as well as keeping the entire system as simple as possible. Data interpretation based on this kind of limited installation will be discussed in section 8.4. However, reducing the number of sensors providing information about the activity of the tenant renders it difficult if not impossible to implement advanced alarm criteria such as identifying pattern changes after the occurrence of a fall or user tracking for the *occupied room criterion*. Depending on the situation and the intended use of an AAL system, it must be decided on a per-case basis whether keeping the number of sensors and thus costs low, i.e., only monitoring the overall activity, or whether numerous redundant sensors shall be installed, thus allowing monitoring the activity of the user much closer and enabling applying more sophisticated alarm criteria than only inactivity monitoring.

7.   Alarm Management

7.1.   Introduction

In the previous chapter, several alarm rules were introduced that can be employed to generate alarms in the first place. However, if alarms are being raised, procedures need to be established that govern how to deal with the raised alarms.

It is of pivotal importance to understand and acknowledge that false alarms are inevitable and cannot be avoided (see 7.3: Necessity of False Alarms). Thus, any attempts to eliminate false alarms will be futile. As a consequence of this realisation, a viable alarm sequence needs to be geared towards handling false alarms – this means, false alarm are not to be penalised but to be seen as integral parts of the entire AAL environment. Little information on the numbers and rates of false alarms that have to be handled by the operators of personal emergency response systems has been published. Yet research is currently being conducted to investigate the implications of false alarms in clinical settings and for emergency call centres in charge of dispatching emergency medical services [Luiz et al., 2000, Kumpch et al., 2010]. The results of these studies, however, cannot directly be transferred to PERS's or AAL systems since both the scope of application and the way of reacting to alarms presented in those two papers and the AAL concept devised in this work overlap only to some extent. The German Red Cross, our partner for handling future alarms generated by future installations of AAL systems, knows from experience, however, that only about 1% of the alarms received by them prove to be genuine medical alerts. Hence, the Red Cross is believed to be ideally prepared as the dispatchers know how to handle the alarms professionally.

In the following, the various steps that will be taken before an internal alarm, i.e., an alarm generated inside a flat by the AAL system, is actually forwarded to an emergency call centre will be elucidated.

7.2.   Alarm Handling and Forwarding

Based on the above introductory reasoning, the alarming scheme devised in this work comprises several steps. The most fundamental, basic case distinction, however, is that the *normal state*, *internal* and *external alarms* are differentiated.

Fig. 7.1 illustrates this fundamental concept. The central control unit (called PAUL, the *p*ersonal *a*ssistive *u*nit for *l*iving, see section 8.1) continuously monitors the inactivity level of the tenant and –if activated– all other alarm criteria defined in previous chapters. As long as all monitored parameters are in their normal ranges, the user and their flat are in the *normal state* and no steps whatsoever need to be taken ("thumb up"). As soon as any of the alarm indicators go out of their permissible range –which may individual to each user–, PAUL will suspect that a potential case of emergency exists ("thumb down"). This can either be caused by genuine emergencies, e.g., if the user becomes unconscious so that all activity ceases, or if PAUL was not able to gather any information about the activity of the user for a certain period of time so that at least one of the alarm thresholds is erroneously exceeded and an alarm is triggered.

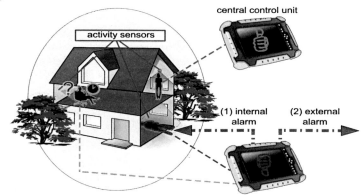

Fig. 7.1: Fundamental alarm principle implemented in the Kaiserslautern and Bexbach pilot projects

Either way, in such a case PAUL will initiate all necessary steps to avert potential dangers or threats from the user. These steps include an internal alarm (stage 2 in the below flow chart) and an external alarm (stage 4).

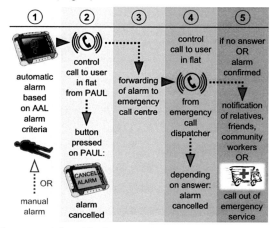

Fig. 7.2: Schematic representation of the five-stage alarm scheme

Fig. 7.2 shows the flow chart representing all steps that will be taken to safeguard the health and safety of the user unless the user cancels an alarm before assistance is administered or an emergency medical service unit is eventually called out.

In step 1, PAUL detects a potential case of emergency. This can have multiple causes: First, the automatic inactivity data processing algorithms running on PAUL may have reasoned that a potential emergency is present. Second, the user may have initiated a manual alarm – either via PAUL or the emergency wall switch in the bathroom. It is, however, inherently impossible for PAUL to differentiate genuine alarms from false alarms: An automatic alarm may be caused by both a *genuine* emergency involving unconsciousness of the user or a *false* alert due to an exceptionally long inactivity phase which is, however, owed to normal user behaviour and not indicative of an emergency. Manual alarms can be raised either because the user did so intentionally by pressing the emergency call button on PAUL or the emergency wall switch in the bathroom or because one of these buttons was pressed inadvertently, e.g., during cleaning or by a visitor who is not familiar with the alarming mechanisms.

In order to intercept as many of these spurious alarms, the alarming scheme comprises step 2: To avoid an overabundance of false alarms being forwarded to the emergency call centre, the user is given the chance to cancel the alarm before it is actually conveyed to the call centre. For this reason, PAUL rings the user on his (mobile) phone. If the user answers the call, a voice message will be played back, notifying the user of the impending alarm forwarding to the emergency call centre. The voice message will as well prompt the user to interact with PAUL within a given short period of time by touching a cancel button displayed simultaneously, allowing the user to stop a potential false alarm from being forwarded. It needs to be noted that solely answering the phone call is not sufficient for cancelling the alarm since an answering machine might also answer the verification call. Making the user press a certain button on the phone to cancel the alarm is also refrained from because it is anticipated that such a call might constitute a somewhat stressful situation for the user in which it may be hard to react properly by means of entering a specific code. The alarm forwarding time delay is typically only a few minutes, giving the user enough time to realise and react to the alarm but not unduly delaying administering help in case of a real emergency. If the alarm does turn out to be a false alarm and the user presses the cancel button displayed on PAUL, the AAL system will switch back to the normal state and continue monitoring the user passively. The interaction with PAUL in order to cancel the false alarm is interpreted as activity so that all inactivity counters are thus reset to zero.

Once step 3 is reached, PAUL assumes that the current alarm is genuine since the user did not cancel it. PAUL now alarms the emergency call centre and transmits all necessary information to the dispatcher, e.g., in which flat the alarm originated, who might be affected, what criterion triggered the alarm, etc. It is important to note, though, that the privacy of the user is ensured to the greatest possible extent since no details regarding the collected raw activity data is transmitted but only the condensed information that at least one alarm criterion went out of the normal range and that assistance may be required immediately.

In step 4, the emergency call dispatcher tries to verify the alarm by calling the user in their flat again. If the user answers the call and informs the dispatcher that a false alarm slipped through, no further actions will be taken, i.e., the alarm is cancelled by the call dispatcher. Stage 5 will not be entered.

In all other cases, i.e., if the phone call does not result in the cancellation of the alarm, the dispatcher proceeds to step 5. There are two conceivable scenarios: First, the user answers but *does confirm* the emergency, e.g., by telling the dispatcher that he fell and now requires help.

Second, the call is not answered at all. In the former case, the operator tries to work out what kind of support is required and dispatches appropriate help. Possible options range from relatives or community workers to medical professionals and emergency medical services (EMS). If the situation cannot be assessed properly, the worst case will be assumed and an emergency unit will be dispatched. The same holds true for the latter case: If the control call is not answered by the user at all, an EMS will as well be dispatched right away.

Even this multi-stage alarm handling scheme, however, will not prevent a certain number of false alarms being reported to the call centre. German emergency call centre operators do know about this fact and are prepared to handle these false alarms professionally. The alarm scheme in this work thus goes far beyond merely accepting or tolerating false alarms. False alarms constitute a vital and important part in the whole scheme and are discussed in detail in the following section.

7.3.   Necessity of False Alarms

As touched upon in previous sections, false alarms are believed not only to be unavoidable but downright indispensable. In this work, they shall substitute mandatory proof tests commonly encountered for instance in chemical industries or process engineering.

Several national and international standards stipulate that process plants must undergo routine inspections (proof tests) at regular intervals. Amongst others, the following standards are concerned with proof tests: VDE/VDI 2180, section 3 [VDIVDE, 2007], DIN EN 61508-4 [DIN, 2002], and DIN EN 61511-1 [DIN, 2005]. In addition, the desirable nature and properties of proof tests are being investigated by numerous researchers worldwide [e.g., [Gabriel et al., 2008, Griffiths, 2006]. In [Gabriel et al., 2008], one of the issues being addressed is spurious trips of safety loops, i.e., triggering of false alarms. In industrial settings, however, spurious trips are undesired because they can cause the plant to go from the normal productive state to a safe state in which the plant is halted. Nevertheless, spurious trips form an integral part of any safety equipment used in process engineering. They are unavoidable and procedures must be put in place to handle spurious trips appropriately. Another aspect that needs to be taken into account in industrial settings is that all safety loops must periodically undergo routine proof tests to ensure that the safety equipment will indeed be capable of fulfilling its task upon a demand, i.e., initiating the transition into a safe state or shutting down the plant. In order to limit the repercussions of such proof tests on normal operation of the plant – namely production interruptions–, the proof test intervals should be as long as possible while at the same time guaranteeing the required safety integrity level SIL (see for example [Aschenbrenner et al., 2005] and [Borcsok, 2007] for fundamentals of safety and reliability engineering). Regarding AAL applications, however, routine checks testing the entire installation from the individual sensors to the emergency call centre are hard to implement and might unduly disturb the user.

The paper by [Griffiths, 2006] has another focus and deals with the implementation of proof testing in software applications. The question that the author addresses is how proof tests can be implemented in systems of which software forms an essential part. The ultimate goal of proof-testing is guaranteeing the dependability of the software. Dependability is an umbrella term comprising several aspects of software engineering among which are availability (i.e., the readiness for usage) and reliability (i.e., "the ability of a system […] to perform its required functions under stated conditions for a specified period of time" [IEEE, 1990]). Nevertheless, proof-testing software turns out to be a very difficult if not downright impossible task. The key issue with software testing is the question as to *what constitutes a proof test*. Since software is assumed to work in a deterministic way, testing it once should suffice. It needs to be noted that two of the main differences between software and physical apparatuses are, on the one hand, that software does not wear out or show signs of deterioration and, on the other hand, that software faults are inherently caused by the design of the software rather than by physical phenomena [Lyu, 1996]. At the same time, there is a plethora of edge conditions –e.g., the operating system, other applications running on the same machine, network load, and many more– that have a potential impact on the operation of the actual software that is to be tested. Since it is very unlikely that all combinations of those edge conditions can be observed or will even occur when testing the software, it cannot be assumed that the behaviour of the software can be tested in its entirety. Thus, several approaches to this issue are proposed in literature: First, proof-testing software can simply mean checking whether the software is operational at a given point in time, i.e., currently working properly. Second, it is proposed to reset (i.e., restart) the software at regular intervals to make sure that for instance memory usage does not become inconsistent. Third, ignoring the fact that one is dealing with software entirely and rather only considering the underlying hardware and its reliability is suggested [Griffiths, 2006]. In addition, four measures are suggested in [Lyu, 1996] to ensure high dependability of software products: fault prevention, fault removal (upon fault localisation), fault tolerance, and fault forecasting. In terms of AAL, however, none of these suggested approaches seems truly satisfactory or even feasible.

Based on the above reasoning, the concept of false alarms substituting proof tests emerged. Just now this idea is also picking up momentum in safety engineering, but to the best of knowledge of the author, no papers have been published on this so far. Using false alarms to verify that the AAL monitoring system does work properly has a number of plausible advantages:

First, a strict software validation and verification process is model-based and would require the existence of models accurately describing the real-world situation [Berard et al., 2001, Litz, 2005]. Since no such models exist, the approach followed in this work was resorting to closely monitoring the system availability by harnessing false alarms as proof tests. Second, false alarms test the *entire* alarming system: By means of a regular proof test, PAUL (the central unit), the software, and the communication line to the emergency call centre could be tested. The operability and availability of the sensors, however, cannot be tested since they work fully automatically. In contrast to that, in case of a false alarm the AAL system does what it is made for – observe a behavioural pattern that is outside the normal range and initiate the alarm sequence. Thus, malfunctions of the system would become apparent if any of its components had failed. Third, moderate numbers of false alarms that do not annoy the user instead train him how to deal with (false) alarms. This contributes to the user feeling confident about the AAL system so that the user knows how to handle and react to the system and alarms in particular. Fourth, false alarms will typically occur much more often than regular proof tests. Thus, the operability of the AAL system will be monitored more closely than in case of normal proof tests. As a result, the requirements towards the individual components constituting the AAL system can be lower than in case of safety equipment tested only once in one or several years. This in turn allows the use of reasonably priced standard components that might otherwise have to be ruled out.

It needs to be ensured, however, that there will not be an overabundance of false alarms that might ultimately turn into an annoyance for the user. To avoid this from happening, the average number of false alarms per month that is acceptable can be limited by the user setting the *mean time between false alarms*. This concept will be discussed in the following section.

7.4.   Mean Time Between False Alarms

As outlined above, false alarms play a vital role in the AAL concept developed in this work. It needs to be ensured, however, that a possibly overly large number of false alarms does not put an undue strain on the users so that they might become irritated or even annoyed. To prevent potentially stressful numbers of false alarms from being raised in the users' flats, the users will be given the opportunity to determine individually and self-determinedly how many days should elapse in between two consecutive false alarms *on average*. Eventually, this results in the concept of the *mean time between false alarms* (MTFA[26]).

> **Definition 7.1: Mean Time Between False Alarms**
>
> The mean time between false alarms MTFA is the number of days having to elapse *on average* in between to false alarms.

By setting the average number of days having to elapse between two false alarms, the user implicitly limits the number of false alarms that may be triggered within a given period of time. It needs to be kept in mind that for example an MTFA of 14 days means three false alarms in an average month. The actual alarms can, however, occur on any of the days. In other words, they can be triggered on the first, the second, and the third day of the respective months just as well as on the first, the fourteenth, and the twenty-eighth day of that month. It should be noted that all considerations relating to MTFA are strictly based on a per-period-of-time approach: E.g., at an MTFA of 14, three false alarms may occur in 28 days since 0000 of day 1 to 2400 of day 14 constitutes a time interval of 14 days as does 2400 of day 14 (= 0000 of day 15) to 2400 of day 28. The following diagrams illustrate the relationship between MTFA and the resulting number of false alarms in 28 days (training period of the monitoring system) and in one year (Fig. 7.3).

---

[26] In the following, *MTFA of n days* and *MFTA n* will be used synonymously.

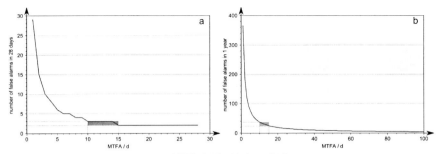

Fig. 7.3: Number of false alarms over MTFA; (a) for 28 days, (b) for one year

From the above two diagrams, it can be concluded that an MTFA of 10–15 days (marked by grey boxes) constitutes a sensible choice. A such chosen MTFA yields reasonable false alarm rates –two or three in a month– which is believed to be acceptable for most users. More frequent false alarms are deemed annoying for the users whereas longer intervals might be detrimental to the desired effect of keeping the users trained in handling the procedures triggered by a false alarm.

Moreover, the MTFA impacts the average response time: The larger the MTFA –and hence the fewer false alarms are accepted–, the more conservatively and less greedy the emergency detection rules must identify possible emergencies. The only way of reducing the number of alarms is raising the alarm threshold, no matter which threshold is applied. As a result of raising the thresholds, the average response time will be elevated as well. The expected correlation between the average response time/the permissible number of false alarms and the MTFA is qualitatively shown in Fig. 7.4.

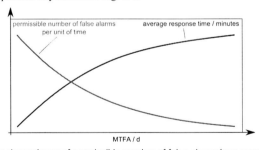

Fig. 7.4: Qualitative dependence of permissible number of false alarms/avg. response time on MTFA

Ultimately, allowing the user to make their own personal choice in terms of the applied MTFA is believed to contribute considerably to the acceptance of the AAL system on part of the user. This is where the MTFA approach comes full circle and takes into account the self-determination of the user stipulated above: The user is free to decide how sensitive the monitoring system should be and may thus trade off disturbance or annoyance possibly felt by false alarms against low response times and quick reactions of the AAL system. Once the user has chosen a specific MTFA, the actual alarm threshold of whatever type being active will be automatically shifted along the ordinate by the monitoring system so that the desired MTFA is yielded.

In essence, the pros and cons of the mean time between false alarms can be summed up as follows:

+ MTFA limits the number of false alarms to a reasonable number, thus preventing undue strain on users.

+ MTFA tests the entire alarming system from sensors and communication channels to the service operator at comparatively short intervals.

+ MTFA trains the users how to deal with (false) alarms.

+ Occasional false alarms make users feel confident that monitoring is indeed working properly.

+ MTFA allows the users to perform trade-off between number of false alarms and average response time of alarming system, thus ensuring self-determination of the users.

+ Ultimately, MTFA is expected to boost the acceptance of the alarming system on the side of the users.

− Alarm thresholds are computed based on previous data. Exactly meeting the desired MTFA in the future cannot be guaranteed.

− Letting the users choose an MTFA autonomously may lead to meaningless settings if the users do not fully understand the effects of their choice.

7.5.   MTFA for Linear Alarm Thresholds

In this section, the recorded inactivity patterns from flats A and B that have already been assessed above will now be subjected to linear alarm thresholds based on various MTFAs. Both 24 hour SLATs and mSLATs only active during certain periods of operation (PO) will be considered.

In case of SLATs, the following two diagrams (Fig. 7.5) clearly illustrate that the MTFA approach is not viable even if allowing false alarms. In flat A, a SLAT of 112 minutes would result if an MTFA of 1 day was chosen. While approximately two hours seem reasonable as an alarm threshold, it must be noted that the vast majority of the resulting alarms occurred at night-time (Fig. 7.6) because the longest inactivity peaks are encountered while the user is asleep. False alarms almost every night are surely not acceptable and certainly detrimental for the overall acceptance of the AAL system. A larger MTFA would not significantly mitigate the problem as the issue regarding the nightly false alarms persisted. On the contrary, increasing the MTFA rather exacerbated the situation since it lead to an elevation of the response time which is not desirable either.

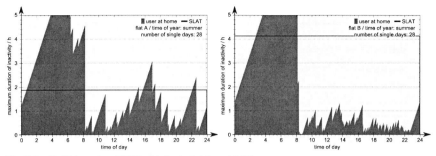

Fig. 7.5: SLAT based on MTFA of 1 day in flat A and flat B

In flat B, the situation is yet worse: When setting an MTFA of 1 day, *all* false alarms occurred during night-time (Fig. 7.6). In addition, a SLAT of more than four hours resulted from that choice for the MTFA. A SLAT this high renders the entire alarming scheme mostly useless. Increasing the MTFA would, on the one hand, reduce the number of false alarms but, on the other hand, lead to an even longer response time while the false alarms were still raised in the middle of the night.

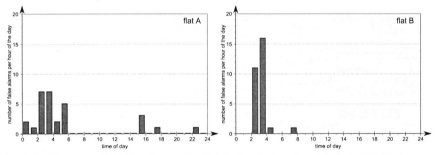

Fig. 7.6: Histograms of the false alarms raised in flats A and B based on data depicted in Fig. 7.5

In order to compare flats in which there is no direct line of sight from the motion detector in the bedroom to the person actually sleeping in the bed, in (Fig. 7.7) the SLAT resulting from an MTFA of 1 day and the accompanying histogram for flat C –in which the person lying in the bed can trigger the motion detector– are shown as well to allow for the comparison of the two different situations regarding the line of sight. The two diagrams illustrate that – even though the overall inactivity at night is lower than in the other flats– the majority of false alarm would still be triggered at night-time. Thus, it has to be concluded as in flats A and B that a 24/7 SLAT is not suitable for raising alarms. It should be noted in addition that due to the overall low inactivity, the SLAT is only about 50 minutes but raising the SLAT would not entirely eliminate false alarms at night-time but mostly avoid them during the day.

In conclusion, all of the above reasoning underscores that the 24/7 SLAT is no appropriate means for raising inactivity alarms. However, with regard to mSLATs as introduced in previous chapters, significant improvements are to be expected when adding the MTFA feature to this type of alarm threshold. Since the period of operation PO was chosen as [0900; 2200] for all previous investigations, this PO will be adhered to in the following as well.

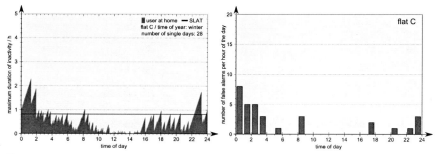

Fig. 7.7: SLAT based on MTFA of 1 day and resulting histogram for flat C

The following diagram (Fig. 7.8) represents the mSLATs$_{09-22}$ that correspond to the various MTFAs. The MTFAs marked by dots in the below diagram were chosen in such a way that all of them yield a different number of permissible false alarms in 28 days (see also Fig. 7.3).

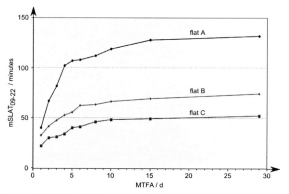

Fig. 7.8: mSLATs$_{09-22}$ over MTFA for flats A, B, and C, as resulting from training data

The empirical evidence displayed in the above diagram corroborates the assumptions made in Fig. 7.4: The lower the MTFA, i.e., the more false alarms per unit of time are tolerated, the lower the response time of the monitoring system can be. However, it turned out that the correlation of mSLAT and MTFA greatly depends on the flat being investigated. While the overall trend is definite, the individual shapes of the curves representing various flats are not of the same dimension. Due to the individual shape of the curves, they represent the typical user behaviour contained in the training data diagrams (see Fig. 7.5 and Fig. 7.7) very accurately: Long inactivity phases during day-time (e.g., as in flat A) yield comparatively high mSLATs whereas low inactivity leads to small mSLATs and thus low response times (e.g., as in flat C).

In order to verify the suitability of the mSLATs in Fig. 7.8 for real-world threshold application, they were applied to the remaining months to determine what number of false alarms they yield in flats A and B in other months than the one used as training data. Since setting the MTFA to values of less than 10 days (i.e., resulting in more than three false alarms per month) or values greater than 15 days (i.e., basically ruling out false alarms on a regular basis) is not assumed to be a reasonable choice, an MTFA of 15 days was chosen for all further investigations. Setting the MTFA to 10 days was refrained from because Fig. 7.8 clearly indicates that there is only a marginal difference in the resulting mSLAT for 10 and 15 days, respectively. However, only for reasons of comparison an MTFA of 29 days was taken into account in the next two tables: Table 7.1 and Table 7.2 list the numbers of false alarms encountered in flats A and B when applying the individual mSLAT to the remaining months[27].

Table 7.1: Numbers of false alarms in flat A applying mSLATs based on MTFAs of 15 and 29 d

Flat A $mSLAT_{09\text{-}22/MTFA15}$ = 128 minutes MTFA = 15 d					Flat A $mSLAT_{09\text{-}22/MTFA29}$ = 132 minutes MTFA = 29 d				
Month	total false alarms	tenant slept longer than usual	failure of monitoring system	genuine false alarms	Month	total false alarms	tenant slept longer than usual	failure of monitoring system	genuine false alarms
S1 2008	2	0	0	2	S1 2008	2	0	0	2
W1 2008	3	0	0	3	W1 2008	2	0	0	2
S1 2009	3	0	0	3	S1 2009	2	0	0	2
S2 2009	2	0	0	2	S2 2009	2	0	0	2

The above table represents the numbers of false alarms generated in flat A when employing the two mSLATs resulting from MTFAs of 15 and 29 days, respectively. In Table 7.2, the same information is listed for flat B.

Fig. 7.9 summarises the results from the above tables. In an average month of 30 or 31 days, an MTFA of 15 days means three false alarms per months on average whereas an MTFA of 29 days corresponds to two false alarms per months on average.

In flat A, applying the MTFA of 15 days to four more months yields two or three false alarms per month which is very close to what is to be expected. In case of the MTFA of 29 days, the predicted number of two false alarms per month is met exactly. These results support the assumption that an MTFA of 15 days or even of 29 days is very suitable for users

---

[27] The data captured in flat C is deemed to be of very good quality. However, only a very limited amount of continuous data is available. Notwithstanding this issue, data from flat C had been used here for demonstration purposes only. Thus, the mSLAT of flat C cannot be verified.

exhibiting activity and inactivity patterns similar to those found in flat A and that the above
MTFAs yield reliable numbers of false alarms.

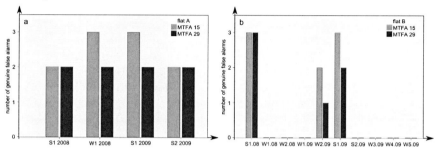

Fig. 7.9: False alarms in flats A and B based on MTFAs of 15 and 29 days

Table 7.2: Numbers of false alarms in flat B applying mSLATs based on MTFAs of 15 and 29 d

	Flat B $mSLAT_{09\text{-}22/MTFA15}$ = 70 minutes MTFA = 15 d					Flat B $mSLAT_{09\text{-}22/MTFA29}$ = 75 minutes MTFA = 29 d			
Month	total false alarms	tenant slept longer than usual	failure of moni- toring system	genuine false alarms	Month	total false alarms	tenant slept longer than usual	failure of moni- toring system	genuine false alarms
S1 2008	3	0	0	3	S1 2008	3	0	0	3
W1 2008	0	0	0	0	W1 2008	0	0	0	0
W2 2008	0	0	0	0	W2 2008	0	0	0	0
W1 2009	0	0	0	0	W1 2009	0	0	0	0
W2 2009	2	0	0	2	W2 2009	1	0	0	1
S1 2009	5	0	2	3	S1 2009	4	0	2	2
S2 2009	0	0	0	0	S2 2009	0	0	0	0
W3 2009	0	0	0	0	W3 2009	0	0	0	0
W4 2009	0	0	0	0	W4 2009	0	0	0	0
W5 2009	0	0	0	0	W5 2009	0	0	0	0

In case of flat B, neither MTFA 15 nor MTFA 29 yield the expected agreement with the predicted numbers of false alarms in the months used for verifying the mSLATs. When interpreting this result, however, two properties of the mSLATs need to be kept in mind: First, both $mSLAT_{09\text{-}22/MTFA15}$ and $mSLAT_{09\text{-}22/MTFA29}$ are rather low –70 and 75 minutes, respectively–, and second, they differ only slightly. This very small difference is reflected by the above diagram – the bar patterns for MTFA 15 and MTFA 29 do not differ substantially. The overall result is, however, that far fewer false alarms would have been raised based on the mSLATs derived from the training data than the expected two and three every month – only in S1 2008, W2 2009, and S1 2009, false alarms would have occurred at all whereas in the other seven months, no false alarms whatsoever would have been triggered. Nevertheless, eight false alarms would have been triggered in ten months when setting the MTFA to 15 days during the training period. This is only about 30% of the predicted number of false alarms, but it must be taken into consideration that the MTFA represents the *mean* time between false alarms and that this value is subject to statistical fluctuations and thus is never to be considered an accurate prediction. However, as shown in Fig. 6.42, the duration of the maximum inactivity peaks in flat B exhibit non-negligible variation that may explain the above finding.

In flat B, it is thus recommended to lower the mSLAT manually if more false alarms are desired. It should be noted, however, that even though the overall activity level in flat B is quite low, lowering the alarm threshold further increases the likelihood that even minor deviations from the daily routine will trigger false alarms.

7.6.   MTFA for Maximum-based Thresholds (MITs)

In this section, the suitability of MITs armed 24/7 is investigated. Other than in the previous section describing mSLATs, the use of MITs aims at monitoring the user round the clock. MITs promise to allow being active at all times because they are based on typical user behaviour and are not simple static linear thresholds. Thus, they take into account the long nightly periods of inactivity and do not trigger an overabundance of false alarms at night-time.

Fig. 7.10 shows sample MITs from flat B. Diagram (a) is based on an MTFA of 10 days and (b) is based on an MTFA of 15 days. In case of mSLATs, no significant difference between MTFA 10 and MTFA 15 could be observed. For that reason, only MTFA 15 was considered in the above case. However, in case of MITs, a substantial drop in the average response time occurs from MTFA 10 to MTFA 15 (Fig. 7.11). Because of that and because of the above reasoning that MTFAs of more than 15 or less than 10 days do not appear reasonable to the author, in the present case of the MITs both MTFA 10 and MTFA 15 were investigated in detail.

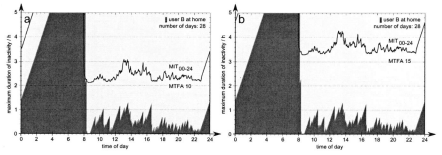

Fig. 7.10: $MIT_{00\text{-}24}$ in flat B, based on MTFA of 10 days (a) and 15 days (b)

The below diagram illustrates the dependence of the average response time $ART(\cdot)$ on the MTFA which in turn determines the MIT. The drop of the $ART(\cdot)$ from MTFA 15 to MTFA 10 is clearly visible. When comparing Fig. 7.8 and Fig. 7.11, it must be kept in mind that mSLATs and ARTs are fundamentally different: An mSLAT is a static linear threshold which is constant throughout the day whereas the ART represents the *average* response time over 24 hours. At night-time, the actual response times are usually much longer than the ART while they can be considerable lower during the day. E.g., in flat B, $ART(MIT_{00\text{-}24/\text{MTFA}10}(\cdot))$ is 244 minutes but this is due to the very long response times at night – Fig. 7.10 (a) illustrates that the response time during the day ranges from approximately 130 to 180 minutes, i.e., is considerable lower and deemed appropriate for the majority of users. The main factor contributing to the actual MIT resulting from a given MTFA is the period during which the user usually gets up – most false alarms will be triggered around that time because, due to the nature of the smoothed MIT, this time span is particularly susceptible to false alarms.

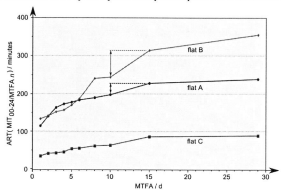

Fig. 7.11: Average response times for various MITs$_{00\text{-}24/\text{MTFA}n}$ over MTFA for flats A, B, and C

The following two tables give the number of false alarms triggered in flats A and B by the respective MITs resulting from MTFA 10 and MTFA 15. The optimum target number of false alarms in an average month is –as discussed above– four for MTFA 10 and three for MTFA 15. Table 7.3 summarises the outcome of the MIT validation in flat A. Several false alarms would have occurred at the time at which the user typically gets up in the morning. This is why the total number of false alarms is rather large – far more false alarms would have been triggered than expected. However, far less "genuine false alarms", i.e., those which occurred at other times of the day than around the getting-up time, were observed (Fig. 7.12 gives a graphical representation of the false alarms). When only genuine false alarms are taken into consideration, the resulting numbers of false alarms are very close to the predicted ones for MTFA 10. In case of MTFA 15, slightly fewer than the expected number of false alarms would have been raised.

Table 7.3: Numbers of false alarms in flat A applying $MIT_{00\text{-}24/MTFA10}(\cdot)$ and $MIT_{00\text{-}24/MTFA15}(\cdot)$

	Flat A $MIT_{00\text{-}24/MTFA10}(\cdot)$ MTFA = 10 d					Flat A $MIT_{00\text{-}24/MTFA15}(\cdot)$ MTFA = 15 d			
Month	total false alarms	getting-up period 0600-0900	failure of moni-toring system	genuine false alarms	Month	total false alarms	get-ting-up period 0600-0900	failure of moni-toring system	genuine false alarms
S1 08	6	4	0	2	S1 08	3	3	0	0
W1 08	8	5	0	3	W1 08	7	5	0	2
S1 09	11	7	0	4	S1 09	5	4	0	1
S2 09	8	3	0	5	S2 09	3	2	0	1

In flat B, the situation is entirely different (see Table 7.4 / Fig. 7.12). Both in case of MTFA 10 and MTFA 15, considerably fewer false alarms than expected were triggered even when taking into account the ones occurring at getting-up time. This finding suggests that the typical behaviour of the user in flat A is fundamentally different from the one of user B. At the same time, the MIT reflects the user much better than the $mSLAT_{09\text{-}22/MTFAxx}$ introduced above: In the above case, only very few alarms were raised at all whereas in case of the MIT, a comparatively low but constant number of alarms is triggered every month – typically one per month both for MTFA 10 and MTFA 15, respectively.

Detailed analysis of the daily rhythm of the user living in flat B showed that the user gets up at a very constant time almost every single morning. Constancy in this particular respect is very favourable for the application of MITs since it leaves little room for false alarms being triggered by vastly varying times of getting up. Hence, it can be concluded that MITs enhanced with MTFAs are a very suitable alarm thresholds for users exhibiting behavioural profiles similar to that of the user living in flat B.

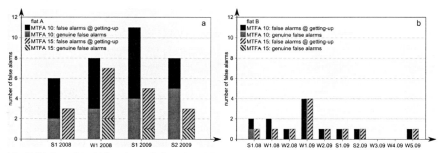

Fig. 7.12: Cumulated false alarms in flats A and B based on (a) $MIT_{0\text{-}24/\text{MTFA10}}(\cdot)$ and (b) $MIT_{0\text{-}24/\text{MTFA10}}(\cdot)$

Table 7.4: Numbers of false alarms in flat B applying $MIT_{00\text{-}24/\text{MTFA10}}(\cdot)$ and $MIT_{00\text{-}24/\text{MTFA15}}(\cdot)$

	Flat B $MIT_{00\text{-}24/\text{MTFA10}}(\cdot)$ MTFA = 10 d					Flat B $MIT_{00\text{-}24/\text{MTFA15}}(\cdot)$ MTFA = 15 d			
Month	total false alarms	tenant slept longer than usual	failure of moni- toring system	genuine false alarms	Month	total false alarms	tenant slept longer than usual	failure of moni- toring system	genuine false alarms
S1 08	2	1	0	1	S1 08	1	1	0	0
W1 08	2	2	0	0	W1 08	1	1	0	0
W2 08	1	1	0	0	W2 08	1	1	0	0
W1 09	4	4	0	0	W1 09	4	4	0	0
W2 09	1	1	0	0	W2 09	1	1	0	0
S1 09	3	1	2	0	S1 09	3	1	2	0
S2 09	1	1	0	0	S2 09	1	1	0	0
W3 09	0	0	0	0	W3 09	0	0	0	0
W4 09	0	0	0	0	W4 09	0	0	0	0
W5 09	1	1	0	0	W5 09	1	1	0	0

Summing up this section, it must be noted that the suitability of the MIT approach is greatly dependant on the particular flat being investigated. For users with inactivity patterns of the type observed in flat B, MITs are a very promising approach to automatic inactivity alarming. However, users exhibiting significant deviations in their daily getting-up time are likely to experience several false alarms per month due to the fact that the MIT is active 24/7 and will easily trip when the user lies in. The number of false alarms is expected to drop, however, if the alarm threshold will be modified in such a way that slight variations in the getting-up time do not trigger an overabundance of false alarms any more.

7.7.   Summary

In the previous chapters, alarming criteria were introduced but the handling and managing of the resulting alarms was left to later consideration. Hence, a suitable alarm handling and management scheme was introduced in the present chapter. The proposed alarm management scheme is capable of dealing with false alarms as well as real alarms and comprises several steps. When an alarm is raised initially, it cannot yet be decided whether or not it is a real or a false alarm. Thus, in order to minimise the number of false alarms actually being forwarded to the emergency call centre, the personal assistive unit for living (PAUL) tries to ring the user before it contacts the call centre. If the user answers the phone, a recorded message will be played back informing the user about the situation and asking him to cancel the alarm if it is a false alarm. Only if the user does not cancel the alarm within a given short period of time, the alarm will indeed be forwarded to the call centre. The emergency dispatcher tries once more to ring the user to investigate what kind of help should be administered or whether a false alarm slipped through. If the user again does not answer at all, a medical emergency unit will be dispatched to determine on-site which situation caused the alarm.

In addition, it was elucidated in detail why false alarms are believed to be indispensable for the successful operation of AAL systems: Since proof tests cannot be conducted at fixed intervals as they might be perceived being intrusive by the user, false alarms shall substitute scheduled proof tests. However, in order for the user to be able to influence to how many false alarms he will be exposed per unit of time on average, the *mean time of false alarms* (MTFA) was introduced. This way, the self-determination of the user can be ensured: Individual users can set their personal MTFA. The MTFA constitutes a trade-off between fast response times in case of a potential health threat and a low number of false alarms.

Finally, the concept of the MTFA was applied to alarming thresholds introduced in previous chapters – SLATs, mSLATs, and MITs. This way, the average response time ART can be lowered because a certain number of false alarms are tolerated. Accepting false alarms is mutually beneficent for both the operator and the user of an AAL system: The system is checked regularly, the user is trained how to handle (false) alarms, and the response time can be lower than if no false alarms were allowed. Simple SLATs have to be ruled out as alarming criteria even when being enhanced by means of the MTFA since most of the false alarms would occur at night which is not deemed acceptable. Avoiding this would require elevating the SLAT so far that the response time is absolutely unacceptable. In contrast to that it was shown that both $mSLAT_{09\text{-}22/\text{MTFAxx}}$ and $MIT_{00\text{-}24/\text{MTFAxx}}(\cdot)$ are appropriate alarming criteria. Their actual suitability, however, greatly depends on the flat under consideration. mSLATs enhanced with MTFA are particularly suitable for flats of type A which are characterised by vastly varying times at which the tenant gets up in the morning and irregular sleeping patterns in general. In spite of these characteristics, the ART was about two hours which is considered acceptable. However, these features may trigger several false alarms if an MIT was to be applied. In flat B, the mSLATs resulted in very low numbers of false alarms even though the response time was only about 70 minutes. In order to generate the desired number of false alarms, the mSLAT would have to be lowered even further manually which in turn rendered it very susceptible even to inactivity peaks of moderate durations. Contrary to that, it was shown that MITs are more favourable in flat B, partly because the time span during which the tenant usually gets up is very narrow and the sleeping patterns are very regular. MITs in flat B yield similar response times during the day as mSLATs in flat A.

8. Practical Implementation

8.1. Introduction

In chapters 4 to 7, novel approaches for supporting persons in need of assistance and safe-guarding their well-being were introduced. The key concept of these approaches is monitoring inactivity since it can be shown that each user exhibits typical behavioural patterns that can be utilised for raising alarms in case of significant deviations of the currently observed inactivity pattern from the learned one. This chapter, however, addresses the real-world implementation of the AAL technology in the users' living environment. The new technology has been installed in two pilot projects in Kaiserslautern and Bexbach, respectively. On top of the actual inactivity monitoring, several other functionalities are provided as well as incentives for the user to accept and use AAL technology on a daily basis.

In the following, an in-depth account of the individual components of the AAL installation will be given: First, PAUL and its functionalities will be described. After that, the two pilot projects will be discussed in detail. Malfunctions encountered in the course of the pilot project are discussed at the end of this chapter.

8.2. The Personal Assistive Unit for Living (PAUL)

8.2.1. Preliminary Remarks

In the introductory chapter, several issues relating to privacy and ethics were touched upon. In this work, one of the fundamental aspects regarding privacy and data security is keeping the raw data of the individual users private. For this reason, the sensor data generated in a flat are stored inside that flat. This can only be accomplished if there is one computer for each flat in which the AAL solution is to be installed. Having a computer in each flat, however, offers numerous advantages. In this work, this computer is called *PAUL*, the *p*ersonal *a*ssistive *u*nit for *l*iving.

First, as mentioned above, an individual computer in each flat allows storing all information –sensor raw data as well as primary, secondary, and tertiary data– locally. Thus, no sensitive private data is disclosed to anyone unless the tenant wishes to participate in a *monitoring and alarming scheme*. Taking part in an alarming scheme is by choice and not compulsory. Moreover, even if an alarm is triggered and the emergency call centre is informed thereof, no details about the sensor data or reasoning having led to this alarm will be shared with the dispatcher. Thus, maximum privacy protection is ensured.

Second, since all monitoring techniques described in previous chapters are represented by algorithms having been implemented as program code, a computer running this software is needed anyway. More importantly, executing the data interpretation and alarm generation algorithms locally on a computer located inside the flat greatly contributes to the protection of the users' privacy as well.

Third, having a computer at one's disposal in each flat an AAL system is installed comes in very handy to give the user the opportunity to control and interact with the AAL system: For instance, visualising the inactivity history along with the inactivity graph of the current day enables the user to monitor their behaviour very easily and to identify possible mismatches between his real activity/inactivity pattern as perceived by the user and the one observed by the AAL system. In addition, being able to review the captured data is believed to improve both acceptance and grasp of the AAL system.

Fourth but not least, the AAL system may in no way stigmatise the users on the basis of their age-related needs or incapabilities. Unfortunately, this is exactly what many well-intentioned products or services offered to seniors do (see also section 3.2.1). On the contrary, if an AAL solution is to be successful, it must provide real benefits for the users which are useful in everyday life without too obviously revealing that age-related deficiencies are existent. These requirements can be met by setting up a sleek computer that is likely to be perceived rather as a cutting-edge gadget than as a clumsy contraption for elderly people. In order to further boost the acceptance and prestige of PAUL and to give the users extra incentives to use it on a daily basis, PAUL has been equipped with additional functionalities: It will assist the user in their everyday life with some of their chores (e.g., rolling up shutters), provide entertainment and means of communication (e.g., Internet access, electronic bulletin forum functionality, web radio), and safety (e.g., video entry phone). These features are believed to be of pivotal importance to motivate users –no matter whether they need health monitoring or not– to use PAUL regularly and thus become acquainted with PAUL, its functionalities, and how to operate it at an age at which doing so does not pose a problem. If PAUL is only installed when prospective users ultimately *need* it, learning how to operate it may form an insurmountable hindrance to the users.

8.2.2.   Hardware Basis for PAUL

As mentioned above, PAUL is a tablet PC featuring a touch screen. In 2003, when the Kaiserslautern AAL project was launched, only few tablet PCs were commercially available that were deemed suitable for the use within the project. Over the course of the project, however, new models were released and taken into consideration for the AAL project. Without delving too deeply into the technical specifications of PAUL, the various models used over time are briefly presented in this section.

The first generation of PAUL of 2006 is displayed in Fig. 8.1. This model had never been deployed to AAL users but only used for internal research and development purposes. At the time when the first coding had been being done, this model was the most suitable one for AAL use. The quality of the speakers and the stand supporting the tablet PC, however, were not satisfactory. Moreover, the enclosure's haptics, colour, and overall appearance were not as appealing as those of PAULv2. Last but not least the price tag of this unit was too high for the technical specifications offered.

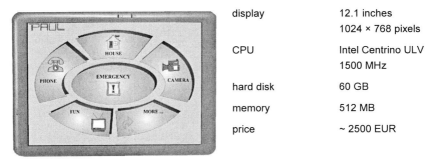

display	12.1 inches
	1024 × 768 pixels
CPU	Intel Centrino ULV
	1500 MHz
hard disk	60 GB
memory	512 MB
price	~ 2500 EUR

Fig. 8.1: First-generation PAUL (PAULv1) – testing model only

Fig. 8.2 shows the second-generation PAUL. Twenty PAULs of this model are being used in the current Kaiserslautern AAL pilot project. In comparison to PAULv1 the price remained the same but CPU power and RAM and hard disk size were improved. The mechanical stability of the enclosure and of the stand is much superior to PAULv1. Moreover, the sound quality also improved significantly which is particularly important for proper functioning of the entertainment functionalities. The basic hardware platform, however, remained the same.

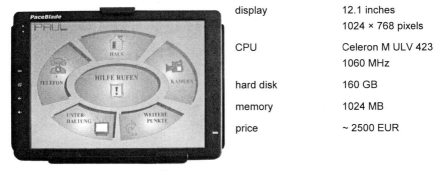

display	12.1 inches
	1024 × 768 pixels
CPU	Celeron M ULV 423
	1060 MHz
hard disk	160 GB
memory	1024 MB
price	~ 2500 EUR

Fig. 8.2: Second-generation PAUL[28] (PAULv2) – currently used in the Kaiserslautern project

In the course of the project it turned out, however, that the fan starts automatically when computationally complex operations, e.g., installing updates, processing large amounts of sensor data, or transmitting data to the university server, are performed. If such operations occur at night, the humming of the fan may disturb the users if PAUL is located in the bedroom (in some of the flats, PAUL is in the sitting room, in other flats PAUL is in the bedroom). This noise was perceived as being very distracting and annoying. Moreover, the small LED indicating that PAUL is powered up also caused discomfort among the users and was reported to keep users from falling asleep. For this reason, and –more importantly– to reduce the entry level price (only 450 EUR for the PC) so that rolling out the novel AAL technology to a wide range of customers becomes feasible, PAULv3 is based on a different hardware

---

[28] Picture courtesy of Gemeinnützige Baugesellschaft Kaiserslautern AG

platform. The model shown in Fig. 8.3 does not come with a battery but is only operational when connected to the mains.

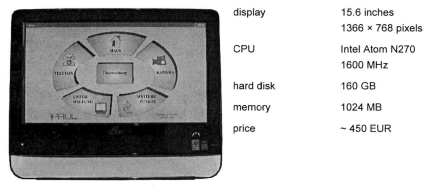

display	15.6 inches
	1366 × 768 pixels
CPU	Intel Atom N270
	1600 MHz
hard disk	160 GB
memory	1024 MB
price	~ 450 EUR

Fig. 8.3: Third-generation PAUL (PAULv3) – currently used in the Bexbach project (German version)

The most important difference to PAULv1 and PAULv2, however, is the scaled-back Atom CPU which is more power-efficient than the fully-fledged CPUs installed in the other versions. On the one hand, this novel low-power CPU causes some loss of computational power but at the same time greatly reduces the power consumed and thus the heat generated during continuous operation. Hence, no fan is needed any more which in turn reduces the noise generated during operation. Regarding the light emission at night, only a small blue LED illuminating the power button is lit at night. PAULv3 is used in the Bexbach pilot project in order to assess whether the greatly reduced computational power of this unit is sufficient for AAL applications in the long run.

8.2.3. Graphical User Interface: Design Principles

PAUL's primary user group are seniors. Thus, the graphical user interface (GUI) needs to be specifically adapted to this target group. The fact that PAUL is not a "conventional" PC but a slim tablet PC featuring a touch screen poses an additional challenge for the GUI design. Already back in the early nineteen nineties, research on the optimum user interface design had been conducted [Lewis & Rieman, 1993-1994, Tognazzini, 1992] and one of the key findings was that "many ideas that are supposed to be good for everybody aren't good for anybody". In other words, the developers of PAUL who are not part of the target user population must keep in mind at all times that seniors in particular shall be able to operate PAUL with ease. For this reason, several fundamental design principles were adhered to:

+ PAUL is to be operated by the touch screen *only*. No keyboard or mouse shall be used or will even be available to the user.
+ The operation concept is to be contrived in such a way that any manuals will be dispensable. I.e., PAUL is self-explanatory.

+ The arrangement of the buttons on the GUI is consistent, i.e., buttons with the same function being displayed on different screens (e.g., the *back* button taking the user back one level) always appear in the same spot. Moreover, one button has exactly one function – the assignment of two or more functions to a single button was refrained from to avoid confusion and unnecessarily complex operation of such buttons.

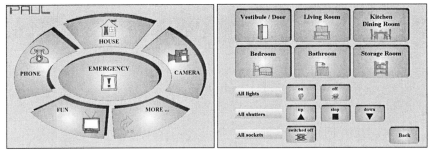

Fig. 8.4: Start screen of PAUL's graphical user interface (left) and *house* sub-menu (right)

+ Moreover, the design of the buttons is consistent as well: All buttons are green with a slight sign of "wear" to resemble physical buttons as closely as possible (see Fig. 8.4). In order to further improve the look and feel of physical buttons, the green buttons displayed on PAUL's GUI seem to recede when being pressed. In addition, the text labels on the buttons are briefly magnified when pressed to acknowledge that the event of pressing the button has been registered by PAUL.

+ Redundant dual representation of all information displayed on the buttons is given in order to facilitate grasp thereof: The information about what purpose a button serves is made accessible both to visual perception and textually. Moreover, the text labels are large enough to be easily legible. The icons are designed to be easy to understand, unambiguous, and of high contrast for effortless perception.

+ In contrast to buttons, non-clickable elements, i.e., informative text boxes are displayed in grey in order to underscore that they are not interactive. Regarding the labelling, the same rules apply as above.

PAUL's graphical user interface, designed in accordance with the above design paradigms, has been in use by about 25 seniors for more than two years at the time of writing this work. User acceptance and the degree to which the design requirements were fulfilled have been evaluated by means of sociological studies. A brief overview of the outcomes of these studies will be given in section 8.6.

8.2.4.   Data Acquisition and Storage

Providing the GUI is not PAUL's only task. On top of that, PAUL receives, stores, and interprets all data telegrams being sent over the KNX bus or transmitted by EnOcean components. Depending on the technology installed in the flats belonging to the two pilot projects, either a KNX gateway (Kaiserslautern) or an EnOcean gateway (Bexbach) link the sensor/actuator array to PAUL. In the former case (KNX), the gateway is bidirectional, i.e.,

PAUL receives all telegrams sent over the bus and is able to send commands to the actuators as well. In the latter case (EnOcean), communication is only unidirectional – PAUL can monitor the telegrams sent by the EnOcean components but cannot send EnOcean telegrams.

All telegrams forwarded by either gateway are delivered to PAUL via the Ethernet interface. Upon reception, the received telegrams are looked up in a table containing all valid telegrams and –based on the publisher/subscriber design pattern [Freeman et al., 2004]– an event is raised that the respective automata needed for processing the telegram are subscribed to. This way, primary, secondary, and tertiary information can be generated. Furthermore, the telegrams are stored in a password protected PostgreSQL database for later transmission to the university's server[29]. For improved data security, the PostgreSQL database is gradually being replaced by proprietary AES256 encrypted files on a per-flat basis.

Moreover, in order allow monitoring whether any problems arise during continuous operation or to facilitate bug tracking, the GUI also adds an "alive" signal every 15 minutes to the database. The alive signals allow assessing retrospectively whether prolonged periods of time during which no telegrams were received represent normal user behaviour (alive signals were created) or were caused by malfunctioning of PAUL (missing KNX/EnOcean telegrams and no alive signals either).

8.2.5.   Additional Functionalities

8.2.5.1.   The Kaiserslautern versus the Bexbach Project

The Kaiserslautern project is by far the more comprehensive project. The flats are equipped with a plenitude of sensors and actuators. In comparison, the Bexbach project is rather to be considered a feasibility study to proof that the concept of a significantly scaled-down approach consisting of only few basic sensors (door sensors and motion detectors) is viable. Such scaled-down installations shall reduce the initial costs of an AAL system and thus improve marketability. However, due to the reduced functional range of the Bexbach AAL solution, several of the below functionalities are only available in the Kaiserslautern installations. Unless stated otherwise, the following statements will represent the Kaiserslautern project.

Moreover, it is important to note that several features of the installed home automation technology installed in the Kaiserslautern AAL-enabled flats serve more than one of the categories comfort, safety, health, and communication/entertainment as listed in Table 3.1 (p. 23). Section 8.3.2 provides details on what kinds and numbers of sensors and actuators the flats are equipped with.

---

[29] All tenants have agreed to this transmission for the purpose of scientific research. See 8.3.3 for details.

8.2.5.2.   Comfort and Home Automation Control

PAUL allows the control of all lights (on/off), windows (open/closed), flat door (open/closed; locked/unlocked), roller blinds (up/down), and the safety power sockets that can be disabled when leaving the flat (see 8.3 for details).

Table 8.1: Home automation functionalities provided by PAUL

	visualise state	actuate individually	actuate en bloc	fine-grained control	aggregate info display
lights	X	X	X		X
windows	X				X
flat door	X	X			
roller blinds	X	X	X	X	X
safety sockets	X	X (only off)[30]			X

Table 8.1 summarises the functions provided by PAUL regarding comfort and home automation. The states of all home automation components can be visualised (e.g., on/off). Apart from the windows, all of these modules can be actuated as well (there are only reed contacts in the window frames but no actuators that could open or close them).

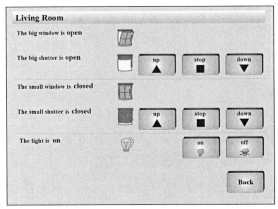

Fig. 8.5: Screenshot of the sitting room controls and visualisation sub-menu

Actuating en bloc means that the entire actuator group can be controlled as a whole, i.e., by pressing a single button, all lights in the flat can be switched on or off which may for example prove useful before going to bed (see Fig. 8.4). *Fine-grained control* means that the

---

[30] For reasons of safety, the safety power sockets may only be switched off remotely but cannot be switched back on using PAUL. Instead, one has to use the wall switches in the kitchen and sitting room, respectively, to enable them again while being able to overlook electrical appliances connected to the sockets.

roller blinds can be actuated independently (see Fig. 8.5) of each other which cannot be accomplished using the wall switches – if there is more than one roller blind in a room, the wall switch always actuates both at the same time, i.e., it is not possible to open or close only a single roller blind. Last but not least, on the *aggregate info display* screen the states of lights, windows, and roller blinds are not displayed on a per-room basis but on an overview screen.

Due to the minimalist approach followed in the Bexbach project, however, none of these functionalities are available there.

8.2.5.3. Safety

The safety aspect comprises two elements: Protecting the tenant and their flat from internal threats (e.g., fire caused by a flat iron) and external threats (e.g., con artists trying to gain access to the flat).

The former class of threats is for example dealt with by means of the switchable power sockets in the sitting room and the kitchen. Flat irons, heaters, or percolators connected to those sockets can be switched of en bloc by touching the respective button on PAUL or by pressing a wall switch next to the flat door. If this button or switch, respectively, is actuated, for example when leaving the flat, all red sockets will conveniently be powered off. Upon the return of the tenant, those red sockets can be switched back on using wall switches in the sitting room or kitchen, respectively. For safety reasons, only sockets may be powered up that can be overlooked in person.

Another mechanism is in place to handle external threats: The block of flat is equipped with a video entry phone. A camera monitors the main entrance. It can be accessed by the tenants through the camera control panel on PAUL (Fig. 8.6). The camera view is displayed on PAUL along with a number of control buttons.

Fig. 8.6: Screenshot of the camera control panel

The tenant can speak to the visitor requesting admission using a bidirectional intercom system that is part of the camera. On the side of the tenant, PAUL's microphone and speakers are used for communication. In addition, a photo of everybody ringing the bell will be taken and stored for 48 hours in the visitors' history. By designing the visitors' history as a ring buffer with a maximum storage time, privacy concerns are accommodated. If the tenant wished to grant someone access to the building, the button *open door* can be pressed.

8.2.5.4.   Health

PAUL is also capable of monitoring the users' health and detecting possible cases of emergency. The algorithms being employed for this were discussed in detail in the previous chapters. The entire monitoring and alarming part of PAUL, however, runs without any human operator being involved. Especially since these monitoring and alarming modules are more or less concealed from the user, they had been presented and discussed in detail at several information sessions for the prospective voluntary users prior to the project launch. It needs to be noted that none of the participants in the pilot projects were pressed or even compelled to accept automatic monitoring but had to opt in. The information sessions were thus conducted in order to elucidate the entire concept as only users who fully understand the AAL technology they are to use can give informed consent thereto. In order to facilitate easy comprehension of the alarm principles in particular, the use of complex, hard-to-understand alarm rules was refrained from so far in the pilot projects. Instead, linear alarm thresholds (mSLATs) only are used for raising inactivity alarms. In order to further stimulate the users' interest in how the automatic monitoring works and to give the users incentives to review the recorded data for defective automatic reasoning, the inactivity graphs can be accessed through PAUL. For reasons of easy comprehension, the alarm thresholds are not automatically established but set by the users. Hence, this section focuses on the way the inactivity is displayed for the users and how they can set the linear alarm threshold manually.

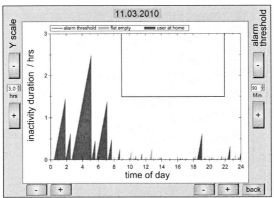

Fig. 8.7: Screenshot of the inactivity visualisation screen

Fig. 8.7 shows the screen that displays the current personal inactivity for a specific user. Phases during which the user is at home or absent, respectively, are marked green and red (dark and light grey in the below figure). The user can change the scaling of the ordinate using the "+" and "–" buttons on the left hand side of the screen.

Since the automatic generation of alarm thresholds suggested is still in beta phase, the current version of the operational end-user version of the inactivity monitoring and alarming software features manual configuration of an mSLAT only. The lower and upper boundary of the mSLAT can be modified using the two sets of "+" / "–" buttons at the bottom of the screen. The actual alarm threshold can be set with the "+" / "–" buttons on the right hand side. As soon as a green (here: grey) spike exceeds the mSLAT, represented by the solid black line, an alarm will be triggered.

In addition, the user can raise an alarm manually by touching the emergency button on the PAUL' start screen (Fig. 8.4). In either case the alarm chain discussed in section 7.2 will be initiated.

Even though the number of activity sensors in Bexbach is smaller than in Kaiserslautern, inactivity monitoring is functional at both locations. In Bexbach, however, activity cannot be monitored as closely as in Kaiserslautern due to the reduced number of sensors.

8.2.5.5. Entertainment/Communication

In order to boost the everyday usefulness of PAUL and thereby elevate the acceptance of AAL technologies on the end of the users, several entertainment and communication functions are provided by PAUL.

The entertainment functions screen is displayed in Fig. 8.8: The users can listen to a number of preset web radio stations (both regional and international ones) or upload MP3 files onto PAUL and listen to them as well.

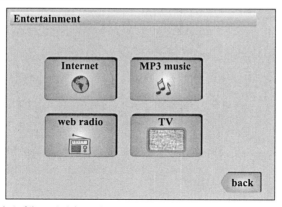

Fig. 8.8: Screenshot of the entertainment menu

Moreover, it is possible to watch cable TV on PAUL. Another feature is an image gallery consisting of photos taken during joint day trips or excursions of the tenants of the block of flat in Kaiserslautern. In the Bexbach project, the tenant can feed their own pictures into that gallery. All of these functions are available both in Kaiserslautern and in Bexbach.

In order to facilitate communication of the tenants of the Kaiserslautern flats with each other, PAUL also features an electronic *bulletin board*. Using a virtual on-screen keyboard, the users can compose short messages (e.g., "Anyone wanting to go to cinema with me tonight?" or "Electric meters will be read next Monday") that will show up on the bulletin board. Other tenants can read and reply to those messages. This function is only available in the Kaiserslautern installation.

8.3.   The Kaiserslautern AAL Project

8.3.1.   The Building and its Tenants

The Kaiserslautern project emerged from a cooperation of the University of Kaiserslautern and a local housing society, Gemeinnützige Baugesellschaft Kaiserslautern AG (BauAG). Fifty-three employees of BauAG administer, maintain, and rent out 5300 flats owned by their company. Over the last years, BauAG went to great efforts to modernise and renovate large parts of their housing stock. One of the buildings that had been modernised is the one shown in Fig. 8.9 and Fig. 8.10, respectively. Before the modernisation, the building looked like in the upper photograph. Since a preservation order had been issued for it, modernising the building was only possible on the condition that the façade of the building (not shown in pictures) was conserved in its original style.

Fig. 8.9: Block of AAL-enabled flats in the Kaiserslautern project *before modernisation*[31]

After a construction phase of one year, the modernised building shown in Fig. 8.10 had been erected. The renovated building accommodates 16 two-room flats, 3 three-room flats, and one single-family home attached to the block of flats. The flats can be accessed from a shared walkway ("access balcony"). The block of flats consists of three storeys and is equipped with a lift to facilitate access to the upper storeys and to ensure handicapped accessibility.

---

[31] Figs. 8.9 and 8.10 by courtesy of Lothar Litz and Gemeinnützige Baugesellschaft Kaiserslautern AG.

Fig. 8.10: Block of AAL-enabled flats in the Kaiserslautern project *after modernisation*

Fig. 8.11 shows a typical ground plot of the two-room flats. The most important sensors used for gathering information about the activity of the tenants are marked in this illustration. The total costs for the sensors, actuators, PAUL, wiring, and installation amounted to approximately 8000 EUR per flat.

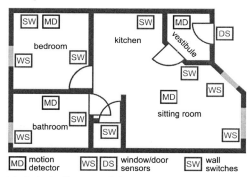

Fig. 8.11: Ground plot of a typical two-room flat

The 26 tenants living in these flats constitute a representative cross-section of the population of Kaiserslautern (as in February 2008): former skilled and unskilled labourers, housewives, as well as two university graduates. Of them, five are 60 years or younger and four are 80 years or older. The majority of tenants is 60 to 80 years old. The mean age is 69 years [Spellerberg et al., 2009].

8.3.2.   Technical Details

8.3.2.1.   Fundamentals of the KNX Bus

All flats were equipped with a KNX (formerly known as EIB) bus during the construction phase. EIB was first introduced in the early nineteen nineties and has a transfer rate of 9600 bits/s. Today, almost all major manufacturers of home automation equipment, white goods,

HVAC (heating, ventilation, and air conditioning) equipment, and related trades offer KNX-compliant products[32].

The advantage of a building automation bus system is that all nodes can exchange data. Hence, by programming the nodes (which essentially are microcontrollers) *inherent intelligence* can be added (e.g., regarding power saving, lighting scenarios, or combined HVAC schemes). Moreover, all components are easily re-programmable so that changing requirements regarding the function of sensors (e.g., light switches) or actuators (e.g., relays) can be accommodated. Finally, the telegrams exchanged by the bus nodes can be tapped for further processing, e.g., using an Ethernet gateway. If a gateway supports bidirectional communication, sophisticated control algorithms can be implemented on an external controller and be fed back onto the bus system [Rose, 1995]. The only drawback is that original KNX equipment is hardly affordable for AAL environments installed by communal housing societies. Thus, instead of a "genuine" KNX bus setup, a centralised approach was followed in the Kaiserslautern project. A simplified schematic representation of the two setups is shown in Fig. 8.12. For reasons of simplicity only lighting is considered here.

If a genuine KNX system is installed, *all* components need to support the protocol natively. For the Kaiserslautern project, this would mean that not only all wall switches but also all lights, window and door sensors, drive shafts of the roller blinds, etc. had to be KNX-compliant which rendered the project unaffordable. In such a setup, the actual electrical appliances and devices were connected to the mains *and* to the KNX bus. Thus, they could be controlled by telegrams sent over the KNX bus. One advantage, however, would be the reduced effort for laying the cables: Both the mains wiring and the bus could be implemented as a ring topology. Sensors as well as actuators could be connected to the bus and the mains anywhere. The KNX bus is, however, independent of the mains grid and no central switchboard is required.

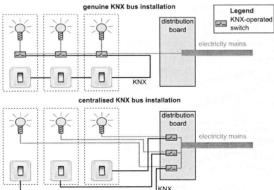

Fig. 8.12: Genuine KNX bus installation in comparison to centralised KNX bus installation

---

[32] Further reading on technical specifications and details of KNX: [Frank, 2009, Rose, 1995]

In contrast to that, the centralised approach mostly relies on electrical standard components available from ordinary ironmongeries: Only the sensors (wall switches for lights/roller blinds and motion detectors) are KNX components. The lamps, drive shafts of the roller blinds, reed sensors for windows and door, etc. are conventional electrical components without KNX support. This necessitates the cables –both mains and KNX– to be laid in a star topology, the centre of the wiring being the distribution board. The distribution board is usually found in an enclosure similar to a fuse box. Several multi-channel KNX relays are mounted on the distribution board which are actuated by the wall switches. For example, if the light is to be turned on in a room, the KNX wall switch is pressed which is connected to the KNX relay on the distribution board. This relay in turn is the actual actuator that switches the light on. Operating the roller blinds works analogously. The reed sensors, however, are only used to determine the states of the windows and the door so that they can be displayed on PAUL or be processed by the finite state machines. They are connected to a multi-channel KNX module converting rising and falling edges of floating contacts (i.e., reed contacts) into KNX telegrams. It needs to be noted that the additional costs for laying both KNX and mains wiring in a star topology are much lower than those for installing KNX components only. However, this is only true for buildings in which cables can be laid during construction phase because extensive retrospective laying of cables is very expensive as well and also requires significant construction work causing noise and dirt. In such a case, wireless solutions are favourable.

In order to allow PAUL accessing the KNX bus for read and write operations, the KNX bus is linked to the in-house Ethernet network to which all PAULs are connected by means of a KNX-Ethernet gateway. In the Kaiserslautern project, a so-called Gira HomeServer 3 (HS3) had been installed[33]. First, the HS3 is a transparent gateway for easy access to the KNX bus. Second, it offers several additional visualisation and automation functionalities that were developed for its use in upmarket private residences.

8.3.2.2.   Sensors and Actuators in the Kaiserslautern Project

Depending on the flat size, a variable number of home automation components had been installed in each of the flats (Table 8.2). All of them are in use for providing comfort, safety, or health functions – most of them serve more than one purpose (Fig. 8.13).

Table 8.2: Types and numbers of home automation components in the Kaiserslautern project

type of component	number
motion detector	4–6
light switch	5–9
roller blind switch	4–8
window sensor (open/closed)	4–8
LED indicating open windows	1
switch for de-energising safety sockets	1
door sensor (open/closed – locked/unlocked)	2
PAUL	1
door opener (via PAUL)	1
TOTAL	**23–37**

---

[33] Manufacturer: Giersiepen GmbH & Co. KG, Germany, see www.gira.de for detailed information.

The below figure illustrates how the multiple sensors contribute to the various AAL base functionality realms. For example, the motion detectors are used to gather information about the users' activity and inactivity, respectively (health and emergency monitoring). At the same time, the motion detectors might be used as one component of a burglar alarm (safety, not currently implemented). Last but not least, they may also be used for switching lights on or off (comfort). True to the maxim of a human-centred, iterative development process and user participation, however, automatic lighting control had been disabled in almost all flats by popular request. The tenants did not wish the lights in their vestibule to go on or off automatically.

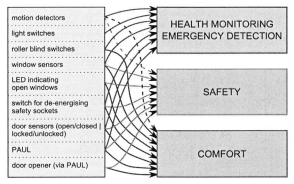

Fig. 8.13: Multiple uses of home automation components for AAL functionalities

Analogous to the motion detectors, all other home automation components contribute to multiple AAL realms as well: Regarding health monitoring and emergency detection, they are crucial for gaining raw activity data. In terms of safety, they assist the tenants in overseeing the state of their flat (i.e., windows being open or closed by means of an LED next to the flat door, visualisation of all components on PAUL). Last but not least, it is simply comfortable to be able to control lights, roller blinds, the door, etc. remotely with PAUL or wall switches.

8.3.3.   Terms and Conditions for the Data Evaluation at the University of Kaiserslautern

The data processing and interpretation steps necessary for health monitoring and emergency detection undertaken at the Institute of Automatic Control at the University of Kaiserslautern have been described and discussed in detail in previous chapters. At this point, however, the terms and conditions for transferring the collected sensor data to as well as storing and processing them at the University of Kaiserslautern are briefly outlined.

The only data collected are the sensor raw data as received by PAUL. All tenants have given their informed consent to a supplement of the tenancy agreement stipulating that named data are transferred to a university's server for the purpose of scientific research in the AAL context. The data is transmitted over an SSL secured connection using the state-of-the-art AES256 cryptographic cipher in conjunction with a cryptographic certificate preventing man-in-the-middle attacks. Access to the server is limited to the author of this thesis only. After having been transmitted onto the server, the data is downloaded from the server at regular intervals –typically once a week– and stored on an encrypted data storage medium. Physical access to that medium is restricted to the author. Moreover, up-to-date cryptographic ciphers (AES, Twofish, Serpent) are used for ensuring data security and access to the encrypted files is limited to the author. In summary, the author has taken –to the best of his knowledge– all reasonable and technically viable measures to prevent unauthorised access to the sensor data and the primary, secondary, and tertiary data derived therefrom.

Similar provisions apply to the Bexbach project. Since the Bexbach house is not yet online, the captured sensor data can only be transferred manually to the university by means of a USB thumb drive. When stored at the university, the data is encrypted using AES256 as well.

8.3.4. Distinctive Features of the Kaiserslautern Project

The Kaiserslautern AAL pilot project differs in several ways from other AAL projects conducted by other research institutes or universities. One of the most prominent features of the project is that it involves long-term tenants living in the block of flats like any other person lives in their home. I.e., the project is not limited in its duration. Moreover, the tenants are "real", i.e., not students or scientists who only conduct short-term tests in the flats. Thus, real long-term everyday user behaviour can be observed.

Moreover, supplementary sociological research has been conducted ever since the project had been launched. The sociological assessment of the project has yielded several valuable results and findings concerning the acceptance of the AAL solution, suggestions for the improvement of PAUL, new functionalities to be implemented, etc.

Last but not least, there are exceptionally strong community bonds among the tenants of the Kaiserslautern block of flats. Although most tenants had not known each other prior to having moved in, an excellent community spirit developed within a very short time. Joint activities include, but are not limited to, coffee parties, day trips in the greater Kaiserslautern area, invited talks on a variety of topics, etc. Moreover, the tenants actively offer help and assistance to each other.

8.3.5. Concluding Remarks

One question not yet specifically addressed is how suitable the presented inactivity monitoring algorithms and the corresponding alarm thresholds derived therefrom are for flats which are inhabited by more than one tenant, e.g., a couple or even a family. After due consideration of this matter, however, it seemed reasonable not to attach too much importance to the issue: There is reason to assume that the occurrence of emergencies or situations that would require automated intervention by an AAL system is by far smaller in flats occupied by more than one person than in case of a person living single. Preliminary research showed that differentiating the contributions of a particular person to the aggregate sensor data captured in a household is a very challenging task. However, if the aggregate sensor signals of more than one person are treated like a homogeneous body of data, it turned out that the same algorithms for deriving representative inactivity patterns and for generating alarms can be employed as in case of a person living single. Future work is expected to shed more light onto this issue.

8.4.   The Bexbach AAL Project

8.4.1.   General Remarks

The Bexbach project was designed as a feasibility study to investigate the suitability of wireless sensors for AAL applications. At the time the Kaiserslautern project was launched, no wireless technologies were mature enough to be used in AAL settings. In the meantime, however, several wireless standards and product lines have emerged. Since –to the best knowledge of the author– the most promising wireless technology is EnOcean[34], only this one had been considered for the Bexbach project. Of all wireless standards known to the author, EnOcean offers by far the most comprehensive product range, including all types of sensors used in the Kaiserslautern project. Moreover, similar to the KNX standard, the EnOcean standard has been adopted by many leading manufacturers of home automation, lighting and HVAC equipment as well as white goods. It is thus to be assumed that long-term support and availability of replacement parts are ensured. Another major advantage of the EnOcean technology is that the sensors are in principle capable of harvesting energy, e.g., by means of solar panels or piezoelectric converters. Thus, no batteries need to be changed and the components are basically maintenance-free. However, under poor lighting conditions, batteries may have to be inserted to back up the capacitor charged by the solar panels.

The reason why wireless AAL solutions are considered being of utmost importance is that only wireless sensors can easily be retrofit into the existing housing stock. If AAL technologies shall ever find widespread use, they must be affordable, quick and easy to install, and may not cause an undue amount of noise or dirt when being installed. Since the vast majority of people lives in their own homes and does not wish to relocate to another dwelling, technologies must be provided whose utilisation in existing homes is feasible. Hard-wired sensors are typically out of the question since installing them would require substantial construction work. Only wireless technologies can fill this gap in the long run. Fig. 8.14 shows a detail of the Bexbach installation. An EnOcean-enabled, solar-powered motion detector is visible in the upper left corner of the door frame. Such a wireless sensor module takes only few minutes to install and is ready for use instantaneously.

In addition to using wireless components, reducing the number of sensors to the absolute minimum for inactivity monitoring is another important aspect of the Bexbach AAL implementation. Since it turned out in the Kaiserslautern project that most of the activity data is captured by the motion detectors rather than wall switches or window sensors, only motion

---

[34] See http://www.enocean.com/ for further information on EnOcean and available products. A detailed discussion of the pros and cons of various wireless technologies would be beyond the scope of this work.

detectors and door sensors (required for determining presence or absence of the tenant) were installed. This way, the initial hardware costs could be cut to about 1500 EUR for three motion detectors, two door sensors, one EnOcean-Ethernet gateway, and one PAULv3. Total installation time for the hardware was half a day.

Fig. 8.14: Interior view of the Bexbach AAL house[35]

As mentioned above, the Bexbach project is a proof-of-concept implementation of the wireless EnOcean technology in an AAL context. As such, close attention was directed to the reliability of the energy harvesting and the wireless communication of the components. The wireless transmission path turned out to be no bottleneck. In the course of the trial operation of EnOcean components, no malfunctions or failures were observed whose cause might have been telegram losses or other radio-communication related malfunctions. Energy harvesting, however, proved to be challenging in existing homes. In winter, when days are short, insufficient amounts of sunlight may reach the solar panels so that continuous operation of the components is not ensured. This issue will be discussed in detail in section 8.5.

8.4.2.   Inactivity Monitoring with the Bexbach Proof-of-Concept Installation

Evaluating the fitness of this scaled-down installation for inactivity monitoring was of particular interest in the Bexbach installation. Thus, PAUL has been recording inactivity profiles for more than one year at the time of writing this to allow comparing these profiles to the ones from Kaiserslautern in order to investigate whether the greatly reduced number of sensors provides sufficient data for reliable inactivity monitoring. Fig. 8.15 shows two sample inactivity graphs representing the user's behaviour in Bexbach. The line of sight from the motion

---

[35] By courtesy of Lothar Litz.

detector in the bedroom to the user lying in the bed is unobstructed. Hence, several brief activities (e.g., turning over while asleep or movement of the extremities) are observed by the motion detector every night. During the day, however, the inactivity level is significantly lower than at night-time. In general, the inactivity patterns obtained in the Bexbach project resemble those from Kaiserslautern very closely. Due to the significantly different triggering and switch-off characteristics of EnOcean motion detectors compared to KNX models (see Fig. 6.33), the inactivity level during the day is even lower than in case of the Kaiserslautern flats discussed above. The below two inactivity diagrams clearly show that basically no inactivity periods of more than only few minutes are observed. The one inactivity spike of more than one hour at 1500 in the right diagram can be attributed to a post-lunch nap. In addition to the ability to monitor inactivity very closely, presence and absence can reliably be differentiated between as well. It can thus be stated that sensor data collected by means of a wireless sensor array only consisting of a motion detector in each room and a door sensor suffices for inactivity monitoring based on the methodologies as introduced using the example of the Kaiserslautern project.

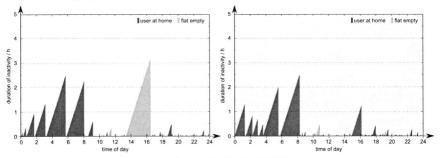

Fig. 8.15: Sample 24 hour inactivity graphs from the Bexbach AAL house

However, it needs to be noted that only continuous activity can be observed using the wireless sensor array described above as no sensors capable of detecting singular events are installed. Thus, more sophisticated alarm rules, e.g., based on separate interpretation of singular events (see 6.4.2), cannot be employed in this basic installation.

In addition to the result that sufficient data for inactivity monitoring could be gathered, it could be proved that the wireless communication link between the sensors and the gateway works very reliably in the Bexbach test setting. No telegram losses were observed as long as sufficient energy is harvested. Telegram losses would have become apparent because EnOcean sensors transmit their status at fixed intervals (typically 1000 seconds).

8.5. Malfunctions and Shortcomings during real Application of the AAL System

When the KNX and EnOcean sensors were first installed in the Kaiserslautern and Bexbach project, severe malfunctions or glitches were not anticipated. It was rather assumed that they function flawlessly since they have been commercially available for several years. It turned out, however, that several types of malfunctions and shortcomings are encountered when using a large number of sensors over long periods of time, thereby increasing the likelihood of encountering such instances.

Two main problem types were identified affecting activity and inactivity monitoring within AAL settings. First, in some cases the technical specifications accompanying commercially available home automation components proved to be insufficient for the full understanding of the mode of operation of the components (e.g., regarding switch-off delay and sensitivity of the motion detectors or the characteristics of the solar panels). Part of the reason for this lack of detailed information is most likely the fact that off-the-shelf home automation components were never designed to be used in AAL settings in which sound knowledge of the components' properties is indispensable. The second field of problems attached to standard automation components are simple malfunctions which again are likely to be caused by the fact that the respective components were not made for mission-critical use in AAL applications. Initially, they were rather made for controlling light or HVAC scenarios. Thus, it is to be assumed that maximum reliability of the components never was a development target.

The below table gives a detailed overview of the issues encountered in the course of the Kaiserslautern and Bexbach AAL project.

Table 8.3: Malfunctions and shortcomings having emerged during practical AAL implementation

motion detectors: spurious signals	Very few of the KNX motion detectors in the Kaiserslautern flats fired at night for no apparent reason. In context with the raw data captured around the time of those spurious trips, there is no reason to assume that there was indeed anyone in the respective room at the given time (e.g., bathroom at night). The actual cause for this behaviour could not yet be determined.
motion detectors: faulty activation by sunlight	At day-time, several motion detectors fired spuriously as well (up to 100 times per day), particularly those in rooms whose windows face north west. Since the observed spurious trips all occurred in the late afternoon, it is assumed that the motion detectors were triggered by the low sun shining onto them. Tests with incandescent lamps in a lab setting showed that the KNX motion detectors are indeed susceptible to changing light intensities and may trigger spuriously due to abrupt changes in intensity.
	However, since the occurrence of this effect is limited to certain weather conditions and was only observed at certain times of the day in only few of the flats, this effect is believed to be negligible. In the worst case, it may trigger a false alarm if the presence FSM erroneously switches back to "tenant at home" because a motion detector fires in a flat several times within a short period of time. If this turns out to be a real problem, a minor change of the presence FSM will stop that behaviour. In the unlikely event that a real case of emergency has occurred and the spurious trips suggest normal activity while there is none, the ORC will cover that issue as well since only bathrooms and bedrooms were affected by these spurious trips.
door sensors: contact bouncing	Two types of sensors are used for monitoring the states of the flat door: A reed sensor determines whether the door is closed or open whereas a lever contact is used to sense whether the door is locked with the lock bolt or unlocked. Either of these two sensors is based on a spring contact that is potentially susceptible to bouncing. In some of the flats, bouncing of the door sensors has been observed. In such a case, several hundred or even thousand door sensor telegrams were received per day, rendering the monitoring system mostly useless because it is then impossible to determine whether the tenant is at home or not. Moreover, the bounce telegrams may suggest activity where there is none. Bouncing, however, did not occur continuously. In some cases, it lasted for a few minutes to hours after the door had been opened, and in other cases it occurred on some days whereas it did not on others.

failure of sensors	In one flat, a bouncing door sensor ultimately failed. To the best of knowledge of the author, this is the only sensor out of more than 600 that became non-operational in two years. Unless it is one of the door sensors which are crucial for determining the users' presence, failures of other sensors can be compensated because of the redundancy that 30+ sensors in a flat provide.
switch-off delays of motion detectors	In case of the KNX motion detectors, the switch-off delay, i.e., the time that has to elapse after the last observed activity before inactivity is signalled, can be set using KNX parameterisation software. In this project, this time is set to the lowest possible value of 12 seconds. In some of the flats, the default value of 300 seconds had not been changed so that the presence FSM did not work correctly as it assumes that the switch-off delay is 12 seconds. Larger values are detrimental for reliable presence/absence detection. This issue could quickly be fixed by means of remote administration.
power supply/ energy harvesting	Initially, EnOcean components were designed for light-flooded functional buildings. When installed in existing houses, solar-powered components may reach their limits. Old houses sometimes only have small windows with comparatively little natural light. Thus, solar panels may not be capable of collecting sufficient energy for continuous operation of the sensors. In lab experiments at the Institute of Automatic Control, it could be shown that the kind of light (incandescent, fluorescent, natural) has a significant impact on the efficiency of the solar panels – natural light and incandescent light are much more effective than fluorescent light. In winter, when only little natural light is available, two out of three motion detectors and two out of two door sensors in the Bexbach flat stopped working. The motion detectors could be equipped with a fallback battery whereas this was not possible in case of the door sensors. Changing the location of the sensor so that it was closer to a window also solved this problem for the respective sensor. Another approach currently being investigated is using artificial lighting to charge the capacitors of the door sensors but evidence gathered so far suggests that this is not a viable option. Thus, a different kind of door sensor must be used, ideally one specifically designed in accordance with AAL demands.
location of motion detectors	Both in the Kaiserslautern flats and the Bexbach house it turned out that the number and positions of the motion detectors are of paramount importance for successfully monitoring activity. Other than initially assumed, one motion detector per room does not suffice for reliably monitoring the tenants' activity in rooms of decent size. I.e., it is well possible that a tenant is a few meters away from the sensor without being detected. In future installations, two motion detectors should be installed in the sitting room to make sure that a good coverage of the room is achieved. In addition, the location where the sensors are mounted is also crucial. This is especially true in the bedroom where the motion detectors are mounted on the wall next to the light switches. In many of the flats, the direct line of sight from the motion detector to the bed is obstructed by cabinets or other furniture. Moreover, the motion detector position at waist-height next to the door frame is far from optimal for achieving a large range of the sensors. In the Bexbach flat, it could be shown that changing the sensor location from the wall opposite the bed to directly above the bed increased the activity rate drastically. Inactivity periods dropped to well below two hours at night.

On top of these technical issues, the degree of compliance and cooperation of the tenants is pivotal for the success of any AAL solution. Since self-determination is of paramount importance in the AAL approach presented in this work, every user has the freedom to switch PAUL off at any time. In order to save electric power, several users switch PAUL off at night or when they are absent for extended periods of time. Hence, no data can be collected at night-time. The times at which PAUL is then switched back on vary from day to day. Thus, these flats are hardly suitable for AAL research due to the lack of consistency of the captured data. Integrating PAUL even more into everyday life (e.g., by including additional features related to comfort, safety, and communication) may help to convince the users to leave PAUL on 24/7.

8.6.   User Acceptance Evaluations in the Kaiserslautern Project

From the onset, the Kaiserslautern pilot project was laid out not to be a field of activity for overly technology-minded engineers but its express purpose was developing viable solutions which both meet the real needs and wants of senior users and can be commercially successful. Thus, the impact of the AAL technology on the seniors making use of it had been continuously assessed throughout the entire project phase. Concomitant sociological research was conducted by Professor Annette Spellerberg and her research group for urban sociology. This chapter only gives a brief overview of these results. For in-depth results please refer to original research papers published by Professor Spellerberg and her research group [Spellerberg et al., 2009, Grauel & Spellerberg, 2008, Grauel & Spellerberg, 2007].

To put it in a nutshell, the outcomes of the sociological research are that the users are well capable of handling and operating PAUL as well as other home automation components. PAUL plays an important role in their everyday life. Over time, the usage intensity has risen rather than stagnated or even decreased which is an excellent result as it underscores that the design concept of the AAL environment was successful.

The interviews with the tenants conducted by Professor Spellerberg and her research group revealed that the tenants are very satisfied with their living condition, the building, and the technology installed in therein. The tenants did not, however, differentiate between the actual *home automation* technology installed in their flats and other features of the block of flats not directly related to AAL, e.g., the lift or the high quality parquet in the flats. All of these aspects equally contributed to their residential satisfaction. Moreover, several tenants could regain independence and mobility because of the excellent location of the building close to the town centre and grocery stores, surgeries, etc. being within walking distance. Last but not least it became clear that most tenants perceive it as a privilege that they were chosen from several hundred applicants for the flats.

Interviews specifically targeting PAUL and the home automation components were conducted two months after the tenants initially moved in and again after another eight months. They showed that the design of the GUI appeals to the users and that the design principles discussed above led to an intuitive and easily comprehensible interface. Legibility of the on-screen texts and size, colour, and contrast of the symbols were judged good. Little tasks (e.g., "open the roller blind in the bedroom") were easily completed. Table 8.4 shows the development of the usage intensity of PAUL.

The below table shows that the overall usage intensity of PAUL increased over time. This can be attributed to the fact that most of the tenants had never used a computer prior to having moved in and were at first reluctant to use PAUL or worried they might break something. However, since PAUL turned out to be a very suitable conversation topic for chats among the new neighbours who did not know each other when moving in, the initial inhibition was soon shed and use of PAUL increased. The by far most often used function of PAUL is the front door camera to check who rang the door bell. Overall, this is a significant success as it is often observed that the initial enthusiasm for technical "gimmicks" wanes quickly. It can thus be stated that after ten months PAUL is used by most of the tenants on a daily basis.

Table 8.4: Usage intensity of PAUL two and ten months after tenants moved into the flats

	users after 2 months	users after 10 months	change
door camera	14	17	+3
Internet	12	10	−2
roller blind control	11	17	+6
light control	1	9	+8
web radio	6	7	+1
alarm clock	3	2	−1

Moreover, the interviews form an integral part of the iterative development process described in chapter 3. Based on the feedback from the users gathered in the interviews, several bugs and flaws of PAUL could be tracked and eliminated: In a small number of flats, lights, windows, or roller blinds were not mapped onto the correct buttons or symbols, e.g., PAUL indicated that the right windows in the bedroom was open when in fact the left one was open. These glitches could be rectified very quickly by means of remote administration. In some other cases, the volume of the speakers had to be adjusted to avoid audio feedback resulting from using PAUL both for listening to web radio and as front door intercom system – turning the volume up caused unacceptable feedback whereas turning the volume down rendered the radio hardly audible. This matter could unfortunately not be resolved in a speedy manner because the audio quality of the built-in microphone and speaker of the camera were far from optimal for the intended use. However, only recently a new generation of IP cameras was released by the manufacturer Mobotix[36] that features a greatly improved microphone, speaker, and audio codec. Another nuisance concordantly reported by the tenants was the automatic lighting of the vestibule: When moving in, the tenants could decide individually whether automatic lighting in the vestibule was desired or not. Of those who opted for it, all had it disabled within the first months of their tenancy because it was perceived annoying. This is also a good example of how people reclaim their self-determination where possible.

In addition to identifying and resolving minor bugs, the interviews also provided valuable feedback as to which enhancements or additional functionalities are desired. In the first generation of the PAUL software, the electronic bulletin board and the photo gallery were not implemented. They were only implemented after the users suggested doing so because they felt that these functions are useful and should be made available.

Last but not least, it must be stated that there are no concerns among the tenants regarding illicit monitoring or surveillance. When prompted about their feelings about inactivity moni-

---

[36] www.mobotix.com

toring by means of the sensors, all tenants showed positive reactions. The tenants approved of the idea of detecting potentially dangerous situations by automatically processing and evaluating the captured sensor data.

8.7.   Summary and Conclusion

In this chapter, the practical implementation of the AAL approach discussed in theory in previous chapters has been described. PAUL, the *personal assistive unit for living*, is the core element of the AAL environment from the users' perspective. PAUL's hardware basis and the GUI are described. In order to give the users incentives to use PAUL routinely and on a daily basis, extra features offering comfort (e.g., remotely controlling the home automation components), safety (e.g., LED warning of windows left open (unintentionally), switching off safety sockets), and communication (e.g., electronic bulletin board, Internet access) were implemented. Those extra features shall motivate the users to overcome possible inhibition thresholds regarding the use of computers.

In the middle part of this section, the Kaiserslautern project, the building, and the technologies used there are discussed in detail. Particular attention is directed to the KNX bus system and its actual implementation in the Kaiserslautern block of flats, the kinds and numbers of sensors and actuators and how these components contribute to more than inactivity monitoring only – most of them are used in realms of safety and comfort as well. Moreover, the terms of service regarding the data transmission to the university are explained. Data security is of paramount importance in this project and all reasonable measures are taken to ensure it.

In comparison to the Kaiserslautern project, the Bexbach project is to be considered a feasibility study rather than a fully-fledged AAL project. The major questions to be answered in this study are discussed (suitability of EnOcean wireless components for AAL usage, reliability of their energy harvesting methods in the existing housing stock, reliability of wireless communication between the nodes, etc.). Even though energy harvesting did turn out to be problematic in winter when days are short, suitable counteractions such as placing the sensors next to a window or equipping them with backup batteries significantly mitigated the power supply problems. Moreover, it is expected that novel wireless components specifically designed for AAL applications and circumventing the mentioned issues will soon be commercially available.

Malfunctions and shortcomings encountered during routine operation of the AAL technology are discussed as well. However, no flaws that could endanger the overall viability of the developed AAL system occurred. Most of the issues observed constitute only minor glitches and can be resolved by optimising the location of the sensor (i.e., no exposure to direct sunlight) or readjusting the sensors (i.e., bouncing door sensors).

In terms of the overall suitability of EnOcean and KNX components for AAL applications, both turned out to be viable options but have specific pros and cons. KNX modules are bus powered so that power supply is not an issue whereas EnOcean components are powered by means of energy harvesting. The latter turned out to be particularly problematic if the components are solar powered (in contrast to, e.g., switches are equipped with piezo units generating power when being pressed). While most of the solar powered modules did become operational by installing them close to a window, rooms without natural light do pose a problem. Future sensor generations are, however, expected to solve this problem by either featuring more efficient solar panels or by supporting optional backup batteries in not enough light can be collected. Regarding the availability of sensors and actuators (wall switches, relays, roller blind actuators, motion detectors, etc.), both KNX and EnOcean offer a wide range of components from a plenitude of manufacturers. The components turned out to be of superior reliability both in terms of failure rates and the communication link. Which technology to choose for a specific installation or purpose ultimately depends on the general conditions of the specific case. As a rule of thumb, it can be said that KNX may be considered in cases in which laying cables is not a problem (e.g., when erecting a new building or completely renovating an existing one), whereas EnOcean is to be preferred in cases in which AAL technology is to be retrofitted into the existing housing stock.

The last section of this chapter addresses acceptance evaluations conducted by Professor Spellerberg. The overall result is that the tenants are very satisfied with the project, enjoy living in the block of flats, use PAUL on a daily basis, and do not feel unduly supervised or monitored.

9.   Summaries

9.1.   Summary

The motivation of this work was developing procedures and methodologies for use in Ambient Assisted Living (AAL) systems. AAL is an umbrella term for a vast array of technologies to help and support (senior) people. However, even though there is a substantial overlap with artificial intelligence, ubiquitous computing, ambient intelligence, and cyberphysical systems, AAL is not to be confused with those concepts. The concept of AAL includes the fields of health monitoring, telecare, emergency detection and avoidance, comfort, safety, and communication. Within the framework of this thesis, health monitoring and automatic alarm generation were of primary interest.

Other than technologies for commercial or industrial applications, AAL solutions are specifically targeted at individuals in their private homes. For this reason, ethical and privacy considerations must be paid particular attention. Hence, an overview of fundamental ethics and privacy considerations was given. In addition, it is pivotal to meet the users' needs and wants, i.e., design solutions and products that indeed do support the users and are appreciated by them. It must be avoided that technology-driven solutions are developed that fail to satisfy the users' expectations. Thus, a strictly iterative development process had been implemented in which several development cycles were undergone. This way, user participation could be ensured and user feedback was valuable input for improving and adjusting the AAL solution according to the users' wishes.

Since this work is largely based on a real-world AAL implementation in Kaiserslautern (Germany), real user data could be collected. The original data that are being captured in the flats, however, are highly unstructured and hardware-dependent. Thus, the raw sensor data are subjected to several data processing steps, yielding primary, secondary, and tertiary data. They are increasingly condensed (i.e., containing less raw sensor data) and more refined. Each of the data processing steps generates more knowledge not directly visible from the raw data, e.g., whether the tenants are at home or not and about the inactivity levels inside a flat.

Inactivity proved to be a suitable means of assessing potential risks or emergencies in a flat. By monitoring user behaviour over a period of four weeks, the typical behavioural patterns were learned by the AAL system and inactivity profiles of each of the participating tenants could be established. Two main procedures for the computation of such profiles were introduced – maximum-based long-term inactivity patterns and long-term inactivity patterns without inactivity peaks that were considered outliers.

Based on these long-term inactivity patterns, alarming principles were developed. Both linear alarm thresholds (i.e., constant throughout the day) and tenant-specific thresholds (i.e., derived from the typical long-term inactivity profile) were used to generate automatic alarms if the current inactivity level in a flat exceeded the respective thresholds. In addition to these general inactivity alarms, advanced criteria were assessed as well: Inactivity pattern changes after a fall has occurred, sojourn times in the various rooms of the flats (occupied room crite-rion ORC), and long-term pattern shifts representing seasonal changes or the onset of chronic illnesses. In conclusion, it turned out that general inactivity thresholds are a viable means of monitoring the health status of the tenants and detecting potential health threats. It needs to be noted, however, that medical emergencies requiring immediate action (e.g., strokes or heart attacks) cannot be detected by the presented monitoring system. Instead, it is very suitable for detecting cases of emergencies in which help needs to be administered within a period of time of minutes to a few hours rather than immediately. The ORC proved to be a useful auxiliary criterion as well. It is best applicable in rooms in which usually short sojourn times are to be expected, e.g., the bathroom. Pattern changes after a fall or long-term pattern shifts were less suited for emergency monitoring. It is believed, however, that this is not a universal truth but that a denser sensor array is likely to render these two criteria very powerful as well.

It is further assumed that in any system raising alarms automatically false alarms cannot be avoided. Thus, procedures were defined and implemented that are capable of handling false alarms. In the presented work, false alarms are not seen as something to be avoided or penalised but rather as an important and valuable tool for giving the user the opportunity to tweak selectivity and response time of the monitoring system. By allowing a small *mean time between false alarms* (MTFA), the system will swiftly report potential emergencies whereas it will react more conservatively if a large MTFA is set. Since false alarms are anticipated in the real-world application of the monitoring system, a multi-stage alarming scheme, involving the users and an emergency response call centre, was introduced and described in detail as well.

The present thesis concluded with an overview of two real-world implementations: In both projects, the *Personal Assistive Unit for Living* (PAUL) is the core element. It is a tablet-PC based device that displays the graphical user interface for the tenants and –at the same time– is the data collection, storage, and interpretation unit. The Kaiserslautern AAL project com-prising 20 flats is equipped with approximately 30 home automation components per flat whereas the Bexbach project is an experimental setup to assess whether low-budget installa-tions consisting of only very few sensors are sufficient for health monitoring. In case of the Kaiserslautern project, the comfort, entertainment, and communication functionalities that had been implemented were illustrated as well.

Last but not least, the author is deeply grateful to the *Ministry of Treasury* of Rhineland-Palatinate (*Ministerium der Finanzen*) and *Stiftung Rheinland-Pfalz für Innovation* (Projekt 961-38 62 61/894) for funding this work. In addition, several partners in the housing sector supported the project as well: *Gemeinnützige Baugesellschaft Kaiserslautern AG, Gemeinnüt-zige Baugenossenschaft Speyer eG, Gemeindliche Siedlungs-Gesellschaft Neuwied mbh*, and *Wohnbau Mainz GmbH*.

Regarding possible future work in this particular field of AAL, the author is of the opinion that the developed data interpretation methods and monitoring technology should be trans-ferred to institutionalised care in order to make it accessible to as many people in need of such technologies as possible. Especially the potential of long-term monitoring of chronic illnesses

Tür- und Fensterkontakten sowie Licht- und Rollladenschaltern (diese Schalter sind „Quasi-Sensoren", da ihre Benutzung ebenfalls Aufschluss über die Aktivität der Bewohner gibt).

Um Informationen über den Zustand der Wohnung oder mögliche Gefahrensituationen für den Bewohner aus den Sensordaten ableiten zu können, werden sie in vier hintereinandergeschalteten Schritten verarbeitet und der Informationsgehalt verdichtet (Abb. 9.1). Zusätzlich wird zusätzliches Wissen durch die Kombination einzelner Sensordaten generiert.

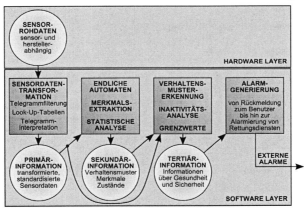

Abb. 9.1: Schritte der Telegrammverarbeitung und Datenauswertung

Schritt 1 umfasst die Standardisierung der Sensordaten. Die sogenannten Sensorrohdaten sind geräte-, hersteller- und systemabhängig. Die entwickelte AAL-Umgebung soll jedoch universell einsetzbar sein und nicht an Sensoren bestimmter Hersteller gebunden sein. Daher erfolgt in diesem Schritt zunächst die Konvertierung der proprietären Datenformate in ein Standardformat (Primärinformationen), das in allen nachfolgenden Schritten einheitlich verwendet wird.

In **Schritt 2** werden die Primärinformationen mit Hilfe von endlichen Automaten und statistischer Analyse weiter verdichtet. Die resultierenden Sekundärinformationen geben Aufschluss über typische Langzeit-Aktivitäts- und -Inaktivitätsmuster der Nutzer sowie über die abgeleiteten Informationen. So existiert z.B. kein Sensor, der direkt die Anwesenheit des Bewohners in der Wohnung ermittelt. Nur durch die Verknüpfung der Daten der Bewegungsmelder mit den Informationen der Türsensoren kann mittels formalisierter Regeln (z.B. als Automat) erschlossen werden, ob in einer Wohnung gerade jemand zu Hause ist oder ob die Wohnung leer ist.

Die **Schritte 3 und 4** beinhalten die Inaktivitätsanalyse und die Alarmgenerierung, die nachfolgend beschrieben werden.

9.2.4.   Inaktivitätsanalyse und Alarmgenerierung

Kernstück der vorliegenden Arbeit ist die automatische Erkennung möglicher Notfälle o-
der anderer Situationen, in denen der Nutzer Hilfe bedarf. Zu diesem Zweck werden aus den
Sensordaten, die in den Wohnungen der jeweiligen Nutzer erfasst werden, innerhalb von 28
Tagen nach der erstmaligen Installation zunächst Referenzinaktivitätsmuster erstellt. Dabei
werden Inaktivitätsphasen, die sich aus der Abwesenheit des Bewohners ergeben, automatisch
durch Regeln erkannt und ausgeblendet. Im Wesentlichen kommen zwei Verfahren zum Einsatz, um derartige Langzeit-Inaktivitätsprofile zu generieren: Das Maximum-Inaktivitätsprofil
bildet zu jedem Zeitpunkt eines Tages die maximal beobachtete Inaktivität während der Lernphase ab, stellt also eine Überlagerung der 28 einzelnen Tagesprofile dar. Dieses Langzeitpro-
fil ist jedoch sehr empfindlich gegenüber einmalig auftretenden, ungewöhnlich langen Inaktivitätsspitzen während der Lernphase. Um dem entgegenzuwirken, wurde das ausreißerberei-
nigte Langzeit-Inaktivitätsprofil eingeführt. Inaktivitätsspitzen, die an einzelnen Tagen auftre-
ten und sich nicht mit den Inaktivitätsprofilen der übrigen Tage der Lernphase decken, werden bei diesem Verfahren ausgeblendet. Somit ist davon auszugehen, dass die ausreißerbereinigten Langzeitprofile das typische Bewohnerverhalten realitätsgetreuer abbilden als die rei-
nen Maximum-Profile. Beide Referenzmuster dienen dazu, für jeden Nutzer individuelle In-
aktivitätsschwellen abzuleiten, die ab dann als Alarmkriterium dienen.

Die Alarmgenerierung stützt sich auf die während der Lernphase generierten typischen Inaktivitätsmuster. Zwei verschiedene Alarmkriterien kommen zum Einsatz: Lineare Alarmgrenzen (*static linear alarm thresholds*, SLAT), die über einen statischen Grenzwert verfü-
gen, und nicht-lineare Alarmgrenzen (*multi-day inactivity-derived threshold*s, MIT), die auf
den ausreißerbereinigten Langzeitprofilen beruhen. Beide Typen weisen spezifische Vor- und
Nachteile auf, die im Folgenden erörtert werden.

Die linearen Alarmgrenzen (Abb. 9.2.a) sind besonders geeignet in Fällen, in denen der
Nutzer insgesamt eine große Aktivität, d.h. geringe Inaktivität, an den Tag legt. Lineare A-
larmgrenzen sind ebenfalls sehr tolerant gegenüber einer großen Variabilität der Inaktivitätsmuster an verschiedenen Tagen, solange die maximal zulässige Inaktivität nicht überschritten
wird. Ein Nachteil der linearen Alarmgrenzen ist, dass sie nachts, wenn der Bewohner schläft,
häufig Fehlalarme aufgrund langer Inaktivität auslösen. Abhilfe verspricht hier die zeitlich
begrenzte Scharfschaltung der Alarmierung (*masked static linear alarm thresholds,* mSLAT)
– so kann z.B. die Zeit von 22 bis 8 Uhr ausgeblendet werden, um eine übermäßige Zahl an
Fehlalarmen in der Nacht zu vermeiden. Um die Toleranz gegenüber Fehlalarmen weiter zu
verbessern, wurde die sogenannte *mean time between false alarms* (MTFA, mittlere Zeit zwischen Fehlalarmen) eingeführt. Mit Hilfe der MTFA kann der Benutzer einstellen, wie viele
Fehlalarme pro Zeiteinheit er zu tolerieren bereit ist. Auf diese Weise wird es dem Benutzer
ermöglicht, selbstbestimmt eine Abwägung zwischen der Zahl der Fehlalarme und der Reaktionszeit des Alarmsystems im Falle eines Notfalls zu treffen – eine große MTFA bedeutet
konservative Alarmierung –d.h. weniger Fehlalarme pro Zeiteinheit– und umgekehrt.

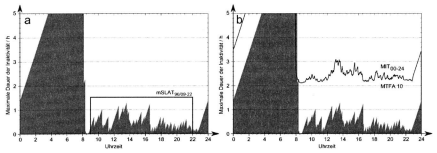

Abb. 9.2: mSLAT und MIT am Beispiel einer 28-Tage-Inaktivitätskurve

Abb. 9.2.b zeigt beispielhaft eine der oben genannten nicht-linearen Alarmgrenzen. Es konnte allerdings gezeigt werden, dass die Anwendung nicht-linearer Alarmgrenzen ohne Berücksichtigung der MTFA nicht praktikabel ist. Mit Ergänzung um die MTFA eignen sich nicht-lineare Alarmgrenzen jedoch ebenso wie die linearen für die Inaktivitätsbeobachtung und Alarmierung (Abb. 9.2.b). Die Stärken der nicht-linearen Alarmgrenzen liegen darin, dass sie 24 Stunden lang aktiv sein können, d.h. auch nächtliche Inaktivitätsphasen abbilden können. Ebenso führen längere Inaktivitätsphasen über den Tag verteilt nicht zu einer Häufung von Fehlalarmen. Allerdings reagieren die nicht-linearen Alarmgrenzen relativ empfindlich auf größere Schwankungen im Tagesrhythmus, was sich besonders in der Zeitspanne, in der der Nutzer typischerweise aufsteht, bemerkbar macht. In zukünftigen Installationen sollte diese besondere Empfindlichkeit noch reduziert werden.

Sowohl im Fall der linearen als auch der nicht-linearen Alarmgrenzen lag die typische Reaktionszeit am Tag (während der Schlafphase war sie höher) bei drei Stunden und weniger. Diese Reaktionszeit ermöglicht zwar nicht die Erkennung von plötzlich auftretenden, lebendbedrohlichen Notfällen, kann jedoch z.B. wirkungsvoll verhindern, dass Menschen stunden- und tagelang am Boden liegen, ohne gefunden zu werden.

Um die Reaktionszeit des Alarmierungssystems unter bestimmten Umständen weiter senken zu können, sind weitere Alarmierungskriterien untersucht worden. Zum einen wurde angenommen, dass sich das Inaktivitätsmuster einer Person z.B. nach einem Sturz ändert – liegt die Person am Boden, ist aber nicht ohnmächtig, kann sie zwar weiterhin Bewegungsmelder auslösen, nicht jedoch Schalter betätigen oder Fenster öffnen. Eine Korrelation der Daten war möglich, jedoch ließen die sich daraus ergebenden Alarmgrenzen keine schnellere Alarmierung als die reine Inaktivitätsüberwachung zu. Zum anderen wurde überprüft, wie lange sich die Bewohner typischerweise am Stück in einem Raum aufhalten, ohne in einen anderen Raum zu wechseln. Es konnte gezeigt werden, dass z.B. im Bad ein Alarm nach 30 bis 60 Minuten gerechtfertigt ist und praktisch keine Fehlalarme auslöst. Im Schlafzimmer oder Wohnzimmer müssen entsprechend höhere Grenzwerte angesetzt werden, die aber dennoch die Inaktivitätserkennung unterstützen können.

Abschließend wurde ermittelt, ob in den über mehrere Monate aufgezeichneten Daten, die für die Untersuchungen zur Verfügung standen, sich ändernde Muster über längere Zeiträume

(Trends) auftraten. Erwartungsgemäß konnten derartige Langzeit-Trends in den vorhandenen Daten jedoch nicht gefunden werden, da die Teilnehmer des Kaiserslauterer Projektes nicht unter chronischen Krankheiten leiden, die zu einer Änderung des Inaktivitätsprofiles hätten führen können. Ferner spielen saisonal bedingte Änderungen der Inaktivitätsprofile ebenfalls nur eine untergeordnete Rolle.

9.2.5. Alarmmanagement

Da das automatische Auslösen von Alarmen im Falle vermuteter Notfälle Ziel der hier beschriebenen Arbeit war, sind Fehlalarme unvermeidlich. Im Laufe der Arbeit wurde klar, dass es nicht sinnvoll ist, Fehlalarme um jeden Preis vermeiden zu wollen, sondern dass sie nützlich sein können und in das Alarmierungskonzept eingebunden werden müssen. Gelegentlich, nicht zu häufig auftretende Fehlalarme testen das gesamte System, stellen sicher, dass es funktionsfähig und verfügbar ist, und zeigen dem Nutzer, dass es tatsächlich im Bedarfsfall einen Notruf auslösen kann. Zudem wird der Nutzer so im Umgang mit (Fehl-)alarmen geschult. Durch die Wahl einer geeigneten MTFA kann der Nutzer die Häufigkeit des Auftretens von Fehlalarmen beeinflussen.

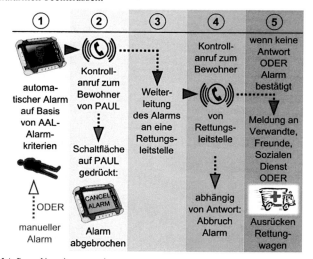

Abb. 9.3: Fünfstufiges Alarmierungsschema

Die eigentliche Alarmierung erfolgt in fünf Schritten (Abb. 9.3). Im ersten Schritt wird entweder ein manueller oder automatischer Alarm ausgelöst. Im zweiten Schritt ruft der *Persönliche Assistent für unterstütztes Leben* (PAUL), der die Benutzerschnittstelle und die Datensammlungs- und Auswerteeinheit verkörpert, den Benutzer an. Bricht der Bewohner den Alarm daraufhin nicht ab, erfolgt die Weiterleitung des Alarms an eine Rettungsleitstelle. In Schritt vier versuchen Mitarbeiter der Leitstelle ebenfalls, den Bewohner telefonisch zu erreichen. Gelingt dies und der Bewohner meldet, dass es sich um einen versehentlichen Fehlalarm handelt, wird der Alarm an dieser Stelle abgebrochen. Nur wenn keine Meldung des Bewohners erfolgt oder er den Alarm bestätigt, werden in Schritt fünf aller erforderlichen Hilfsmaß-nahmen getroffen. Das kann unter anderem eine Meldung an Angehörige oder Freunde sein oder auch das Aussenden eines Rettungswagens.

9.2.6. Praktische Umsetzung

Die vorgestellte Arbeit ist aus einem realen AAL-Projekt in Kaiserslautern, das in Zu-sammenarbeit der TU Kaiserslautern und der Gemeinnützigen Baugesellschaft Kaiserslautern AG (BauAG), realisiert wurde, hervorgegangen. Im Jahr 2007 hat die BauAG eine ihrer Wohnanlagen von Grund auf saniert und dabei rund 20 Wohnungen mit gehobener, kabelge-bundener KNX-Hausautomatisierungstechnik ausgestattet.

Kernstück der AAL-Technik ist PAUL, der in jeder der Wohnungen einmal vorhanden ist. Mit seiner Hilfe können die Bewohner Rollläden und Lichter steuern, auf die Haustürkamera zugreifen, auf das Internet und Internetradio zugreifen, ein elektronisches schwarzes Brett nutzen und vieles andere mehr. Diese Komfort-, Sicherheits- und Kommunikationsfunktionen sollen den Nutzen der AAL-Installation vergrößern und Anreiz zur möglichst häufigen Nut-zung des Systems sein.

Die Akzeptanz und der Nutzen der entwickelten AAL-Lösung wurden von Prof. Annette Spellerberg sozialwissenschaftlich untersucht. Die Ergebnisse dieser Untersuchungen sind auf Konferenzen und in Fachjournalen veröffentlicht [Grauel & Spellerberg, 2007, Grauel & Spellerberg, 2008, Spellerberg et al., 2009]. Zusammenfassend ist festzustellen, dass die Wohnzufriedenheit der Bewohner des Kaiserslauterer AAL-Projektes ausgezeichnet ist. E-benso ist hervorzuheben, dass von Seiten der Bewohner keine Angst vor Überwachung oder Verletzung der Privatsphäre besteht.

In einem weiteren, kleineren Pilotprojekt in Bexbach wurde ein Haus im Wohnbestand mit funkbasierter EnOcean-Hausautomatisierungstechnik ausgestattet. Kabelgebundene Systeme sind im Wohnbestand praktisch nicht nachrüstbar. Ziel dieses Bexbacher Pilotprojektes war es, die zuverlässige Funktion von Funkkomponenten zu zeigen und gleichzeitig durch den Einsatz möglichst weniger Module den Einstiegspreis zu minimieren, um so den möglichst breiten Einsatz von AAL-Technik gewährleisten zu können.

9.2.7. Ausblick

Der Autor ist bestrebt, die im Rahmen dieser Arbeit entwickelten und vorgestellten Ver-fahren zur Inaktivitätserkennung und Alarmgenerierung auf Anwendungen im Pflegebereich zu übertragen. Besonders im Hinblick auf chronische Erkrankungen erscheint die kontinuier-liche Erfassung von Sensordaten erfolgversprechend, um Beginn und Verlauf chronischer Krankheiten erkennen und verfolgen zu können. Die gewonnenen Erfahrungen in Bezug auf geeignete Sensortechnologien; Zahl, Art, Platzierung der verwendeten Sensoren; sowie die Methoden und Verfahren zur Datenauswertung und –interpretation bilden dabei eine wertvol-le Grundlage. Es wird erwartet, dass derartige AAL-Systeme im professionellen Pflegeumfeld zu einer Entlastung der Pflegekräfte und Ärzte von Routineaufgaben führen wird und diese somit mehr Zeit für die eigentliche Betreuung der Patienten haben werden.

10.   Bibliography and Indices

10.1.   Bibliography

[Adam, 2009] ADAM, CLEMENS. 2009. *Hausnotruf. Mehr Lebensqualität und Sicherheit - jederzeit!* Tech. rept.
Ministerium für Arbeit, Gesundheit und Soziales des Landes Nordrhein-Westfalen.

[AIHW, 2008] AIHW. 2008. *Indicators for chronic diseases and their determinants 2008.* Australian Institute of
Health and Welfare (AIHW).

[Allhoff *et al.*, 2009] ALLHOFF, FRITZ, LIN, PATRICK, MOOR, JAMES, & WECKERT, JOHN. 2009. *Ethics of Human
Enhancement: 25 Questions & Answers.* Tech. rept. US National Science Foundation.

[Alwan *et al.*, 2006]ALWAN, M., DALAL, S., MACK, D., KELL, S. W., TURNER, B., LEACHTENAUER, J., & FELDER,
R. 2006. Impact of monitoring technology in assisted living: Outcome pilot. *IEEE Transactions On
Information Technology In Biomedicine,* **10**(1), 192–198.

[amsys, 2010] AMSYS. 2010. *Ambient Systems - Technologies and Applications.* Online. http://www.amsys-uni-
kl.de/index.php?L=en&ID=start&NAV=0 (as of 13 May 2010).

[Anliker *et al.*, 2004] ANLIKER, URS, WARD, JAMIE A., LUKOWICZ, PAUL, TROESTER, GERHARD, DOLVECK,
FRANCOIS, BAER, MICHEL, KEITA, FATOU, SCHENKER, ERAN B., CATARSI, FABRIZIO, COLUCCINI, LU-
CA, BELARDINELLI, ANDREA, SHKLARSKI, DROR, ALON, MENACHEM, HIRT, ETIENNE, SCHMID, ROLF,
& VUSKOVIC, MILICA. 2004. AMON: a wearable multiparameter medical monitoring and alert sys-
tem. *IEEE Trans Inf Technol Biomed,* **8**(4), 415–427.

[Aschenbrenner *et al.*, 2005] ASCHENBRENNER, S. H., FALLER, R. I., GOBLE, W. M., GREBE, J. C., OPEM, A., VAN
BEURDEN, I. J. W. R. J., & VAN BEURDEN-AMKREUTZ, R. J. P. 2005. *Safety equipment reliability
handbook.* EXIDA.

[Ayari & Tielert, 2007] AYARI, E., & TIELERT, R. 2007. Front End Schaltung zur Online Auswertung von EKG-
Signalen. *Advances in Radio Science,* **5**(June), 197–204.

[Barger *et al.*, 2005] BARGER, T. S., BROWN, D. E., & ALWAN, M. 2005. Health-status monitoring through
analysis of behavioral patterns. *IEEE Transactions On Systems Man And Cybernetics Part A-Systems
And Humans,* **35**(1), 22–27.

[Barnes *et al.*, 1998] BARNES, N. M., EDWARDS, N. H., ROSE, D. A. D., & GARNER, P. 1998. Lifestyle monitoring - technology for supported independence. *Computing & Control Engineering Journal,* **9**(4), 169–
174.

[Bechtold & Sotoudeh, 2009] BECHTOLD, ULRIKE, & SOTOUDEH, MAHSHID. 2009. Ambient Assisted Living
(AAL) as a promoter of the good life – some conditions/aspects from a technology assessment perspective. *Pages 188–189 of: Proceedings of SPT 2009 - Converging Technologies, Changing Societies.*

[Beckmann *et al.*, 2004] BECKMANN, C., CONSOLVO, S., & LAMARCA, A. 2004. Some assembly required: Supporting end-user sensor installation in domestic ubiquitous computing environments. *Ubicomp 2004:
Ubiquitous Computing, Proceedings,* **3205**, 107–124.

[Becks *et al.*, 2007] BECKS, THOMAS, DEHM, JOHANNES, & EBERHARD, BIRGID. 2007. *Ambient Assisted Living -
Neue "intelligente" Assistenzsysteme für Prävention, Homecare und Pflege.* Tech. rept. Deutsche Ge-
sellschaft für Biomedizinische Technik im VDE.

[Berard *et al.*, 2001]   BERARD, B., BIDOIT, M., FINKEL, A., LAROUSSINIE, F., PETIT, A., PETRUCCI, L., & SCHNOEBELEN, P. 2001. *Systems and software verification: model-checking techniques and tools.* New York, NY, USA: Springer-Verlag New York, Inc.

[Boehm *et al.*, 2005]   BOEHM, BARRY, ROMBACH, HANS DIETER, & ZELKOWITZ, MARVIN V. 2005. *Foundations of Empirical Software Engineering: The Legacy of Victor R. Basili.* Secaucus, NJ, USA: Springer-Verlag New York, Inc.

[Borcsok, 2007]   BÖRCSÖK, JOSEF. 2007. *Elektronische Sicherheitssysteme. Hardwarekonzepte, Modelle und Berechnung.* Huethig.

[Borges *et al.*, 2008]   BORGES, I., SINCLAIR, D., MOLLENKOPF, H., RAYNER, P., BOND, R., & PARENT, A.-S. 2008 (November). *Older people and Information and Communication Technologies - An Ethical approach.* Online. http://www.age-platform.org/EN/IMG/pdf_AGE-ethic_A4-final.pdf (as of 14 Aug 2009).

[Bostrom & Sandberg, 2009]BOSTROM, N., & SANDBERG, A. 2009. Cognitive Enhancement: Methods, Ethics, Regulatory Challenges. *Science and Engineering Ethics*, **15**(3), 311–341.

[Bostrom & Roache, 2007]   BOSTROM, NICK, & ROACHE, REBECCA. 2007. *Ethical Issues in Human Enhancement.* Palgrave Macmillan. Chap. Human Enhancement.

[Brant, 1990]BRANT, ROLLIN. 1990. Comparing Classical and Resistant Outlier Rules. *Journal of the American Statistical Association*, **85**(412), 1083–1090.

[Brennan & Cardinali, 2000]BRENNAN, M., & CARDINALI, G. 2000. The use of preexisting and novel coping strategies in adapting to age-related vision loss. *Gerontologist*, **40**(3), 327–334.

[Bruns *et al.*, 1999]   BRUNS, UTA, HILBERT, JOSEF, & SCHARFENORTH, KARIN. 1999. Technik und Dienstleistungen für mehr Lebensqualität im Alter: Ansätze zur Gestaltungsorientierung. *Zeitschrift für Sozialreform*, **45**, 369–382.

[Cook, 2006]COOK, D.J. 2006. Health Monitoring and Assistance to Support Aging in Place. *Journal of Universal Computer Science*, **12**(1), 15–29. |http://www.jucs.org/jucs_12_1/health_monitoring_and_assistance|.

[Dalal *et al.*, 2005]   DALAL, SIDDHARTH, ALWAN, MAJD, SEIFRAFI, REZA, KELL, STEVE, & BROWN, DONALD. 2005. A Rule-Based Approach to the Analysis of Elders' Activity Data: Detection of Health and Possible Emergency Conditions. *In: AAAI 2005 Fall Symposium.*

[Datta, 2001]DATTA, BISWA NATH. 2001. *Applied and Computational Control, Signals, and Circuits.* Springer, Berlin.

[DIN, 2002] DIN. 2002. *DIN EN 61508-4: Funktionale Sicherheit sicherheitsbezogener elektrischer/elektronischer/programmierbar elektronischer Systeme - Teil 4: Begriffe und Abkürzungen.*

[DIN, 2005] DIN. 2005. *DIN EN 61511-1: Funktionale Sicherheit - Sicherheitstechnische Systeme für die Prozessindustrie - Teil 1: Allgemeines, Begriffe, Anforderungen an Systeme, Software und Hardware.*

[Ekstedt *et al.*, 2006]   EKSTEDT, MIRJAM, SODERSTROM, MARIE, AKERSTEDT, TORBJORN, NILSSON, JENS, SONDERGAARD, HANS-PETER, & ALEKSANDER, PERSKI. 2006. Disturbed sleep and fatigue in occupational burnout. *Scand J Work Environ Health*, **32**(2), 121–31.

[EU, 2004]EU. 2004. *Ambient Intelligence.* Online. http://ec.europa.eu/information_society/tl/policy/ambienti/index_en.htm (as of 2 Nov 2009).

[EU, 2006]EU. 2006. *The Ambient Assisted Living (AAL) Joint Programme.* Online. http://ec.europa.eu/information_society/activities/einclusion/research/docs/aal_overview_16_june_2008.pdf (as of 28 Oct 2009).

[EU, 2008]EU. 2008. *Ambient Assisted Living (AAL) Joint Programme - Call for Proposals 2008 - AAL-2008-1 -*
"ICT based solutions for Prevention and Management of Chronic Conditions of Elderly People". Online. https://ssl.aal-europe.eu/calls/aal-2008-1 (as of 28 Oct 2009).

[Eurostat, 2008] EUROSTAT. 2008. *Vorausgeschätzter Altersquotient*. Online. http://epp.eurostat.ec.europa.eu/
tgm/table.do?tab=table&init=1&plugin=1&language=de&pcode=tsdde511 (as of 22 Oct 2009).

[Eurostat, 2009a] EUROSTAT. 2009a. *Bevölkerungsprognosen*. Online. http://epp.eurostat.ec.europa.eu/tgm/
table.do?tab=table&init=1&plugin=1&language=de&pcode=tps00002 (as of 22 Oct 2009).

[Eurostat, 2009b] EUROSTAT. 2009b. *Lebenserwartung bei der Geburt, nach Geschlecht (Jahre)*. Online.
http://epp.eurostat.ec.europa.eu/tgm/refreshTableAction.do?tab=table&plugin=1&init=1&pcode=tps0
0025&language=de (as of 22 Oct 2009).

[Ferre & Medinilla, 2007] FERRE, XAVIER, & MEDINILLA, NELSON. 2007. How a Human-Centered Approach
Impacts Software Development. *Pages 68–77 of: Human-Computer Interaction. Interaction Design*
and Usability. Lecture Notes in Computer Science, vol. 4550. Springer Berlin / Heidelberg.

[Floeck & Litz, 2007] FLOECK, MARTIN, & LITZ, LOTHAR. 2007. Ageing in Place: Supporting Senior Citizens'
Independence with Ambient Assistive Living Technology. *mst|news*, **6/2007**(6), 34–35.

[Floeck & Litz, 2008a] FLOECK, MARTIN, & LITZ, LOTHAR. 2008a. Activity- and Inactivity-Based Approaches
to Analyze an Assisted Living Environment. *Pages 311–316 of: Proc. Second International Confer-*
ence on Emerging Security Information, Systems and Technologies SECURWARE '08.

[Floeck & Litz, 2008b] FLOECK, MARTIN, & LITZ, LOTHAR. 2008b. Integration of Home Automation Technology into an Assisted Living Concept. *In:* KARSHMER, ARTHUR I., NEHMER, JÜRGEN, RAFFLER, HART-
MUT, & TROESTER, GERHARD (eds), *Assisted Living Systems - Models, Architectures and Engineering*
Approaches. Dagstuhl Seminar Proceedings, no. 07462. Internationales Begegnungs- und For-
schungszentrum für Informatik (IBFI), Schloss Dagstuhl, Germany.

[Floeck & Litz, 2009] FLOECK, MARTIN, & LITZ, LOTHAR. 2009. Inactivity Patterns and Alarm Generation in
Senior Citizens' Houses. *Pages 3725–3730 of: Proc. European Control Conference 2009 Budapest*.

[Fornara *et al.*, 2001] FORNARA, P., DOEHN, C., FRESE, R., & JOCHAM, D. 2001. Laparoscopic nephrectomy in
young-old, old-old, and oldest-old adults. *J Gerontol A Biol Sci Med Sci*, **56**(5), M287–M291.

[Frank, 2009] FRANK, KARLHEINZ. 2009. *EIB/KNX Grundlagen Gebäudesystemtechnik*. Verlag Technik /Huss
Media.

[Freeman *et al.*, 2004]FREEMAN, ELISABETH, FREEMAN, ERIC, BATES, BERT, & SIERRA, KATHY. 2004. *Head*
First Design Patterns. 1 edn. O'Reilly Media.

[Fugger *et al.*, 2007] FUGGER, E., PRAZAK, B., HANKE, S., & WASSERTHEURER, S. 2007. Requirements and
ethical issues for sensor-augmented environments in elderly care. *Universal Access in Human Com-*
puter Interaction: Coping with Diversity, Pt 1, **4554**, 887–893.

[Gabriel *et al.*, 2008] GABRIEL, THOMAS, LITZ, LOTHAR, & SCHROERS, BERND. 2008. A formal approach to
derive configurable Markov models for arbitrarily structured safety loops. *In: Proceedings of the 9th*
International Conference on Probabilistic Safety Assessment and Management (PSAM9), Paper ID
132.

[Gardner, 1985] GARDNER, EVERETTE S. 1985. Exponential Smoothing: The State of the Art. *Journal of Fore-*
casting, **4**, 1–28.

[Georgieff, 2008]GEORGIEFF, PETER. 2008. *Ambient Assisted Living - Marktpotenziale IT-unterstützter Pflege*
für ein selbstbestimmtes Altern. FAZIT Schriftenreihe, no. 17. MFG Stiftung Baden-Württemberg.

[Glascock *et al.*, 2007] GLASCOCK, ANTHONY P., & KUTZIK, DAVID M. 2007. An Evidentiary Study of the Uses of Automated Behavioral Monitoring. *Pages 858–862 of:* KUTZIK, DAVID M. (ed), *Proc. 21st International Conference on Advanced Information Networking and Applications Workshops AINAW '07*, vol. 2.

[Grauel & Spellerberg, 2007] GRAUEL, JONAS, & SPELLERBERG, ANNETTE. 2007. Akzeptanz neuer Wohntechniken für ein selbstständiges Leben im Alter - Erklärung anhand sozialstruktureller Merkmale, Technikkompetenz und Technikeinstellungen. *Zeitschrift für Sozialreform*, **2**, 191–215.

[Grauel & Spellerberg, 2008] GRAUEL, JONAS, & SPELLERBERG, ANNETTE. 2008. *Attitudes and Requirements of Elderly People Towards Assisted Living Solutions*. Heidelberg: Springer.

[Griffiths, 2006] GRIFFITHS, ALENA. 2006. On proof-test intervals for safety functions implemented in software. *Pages 23–33 of: SCS '06: Proceedings of the eleventh Australian workshop on Safety critical systems and software*. Darlinghurst, Australia, Australia: Australian Computer Society, Inc.

[Grubbs, 1950] GRUBBS, FRANK E. 1950. Sample Criteria for Testing Outlying Observations. *The Annals of Mathematical Statistics*, **21**(1), 27–58.

[Hadidi & Noury, 2009] HADIDI, TAREQ, & NOURY, NORBERT. 2009. A Predictive Analysis of the Night-Day Activities Level of Older Patient in a Health Smart Home. *Pages 290–293 of: ICOST '09: Proceedings of the 7th International Conference on Smart Homes and Health Telematics*. Berlin, Heidelberg: Springer-Verlag.

[Haigh *et al.*, 2006] HAIGH, K.Z., KIFF, L.M., & HO, G. 2006. The Independent LifeStyle Assistant (I.L.S.A.): Lessons Learned. *Assistive Technology*, **18**, 87–106.

[Hughes & Louw, 2002] HUGHES, JULIAN C, & LOUW, STEPHEN J. 2002. Electronic tagging of people with dementia who wander. *BMJ*, **325**(7369), 847–848.

[IEEE, 1990] IEEE. 1990 (Dec.). *IEEE Standard Glossary of Software Engineering Terminology 610.12-1990*.

[Jones, 2007] JONES, M. TIM. 2007. *Artificial Intelligence: A Systems Approach*. Infinity Science Press.

[Kaufman *et al.*, 2003] KAUFMAN, D. R., PATEL, V. L., HILLIMAN, C., MORIN, P. C., PEVZNER, J., WEINSTOCK, R. S., GOLAND, R., SHEA, S., & STARREN, J. 2003. Usability in the real world: assessing medical information technologies in patients' homes. *Journal Of Biomedical Informatics*, **36**(1-2), 45–60.

[Kimpeler *et al.*, 2006] KIMPELER, SIMONE, BAIER, ELISABETH, DOEBLER, THOMAS, KIRCHMAIR, ROLF, DOH, MICHAEL, MOLLENKOPF, HEIDRUN, GAUGISCH, PETRA, KLEIN, BARBARA, & SCHMIDT, MARTIN. 2006. "Best Agers" in der Informationsgesellschaft. *In:* KIMPELER, SIMONE, & BAIER, ELISABETH (eds), *IT-basierte Produkte und Dienste für ältere Menschen - Nutzeranforderungen und Techniktrends*.

[Kleinberger *et al.*, 2009] KLEINBERGER, THOMAS, JEDLITSCHKA, ANDREAS, STORF, HOLGER, STEINBACH-NORDMANN, SILKE, & PRUECKNER, STEPHAN. 2009. An Approach to and Evaluations of Assisted Living Systems Using Ambient Intelligence for Emergency Monitoring and Prevention. *Pages 199–208 of: HCI (6)*.

[KNX, 2006] KNX, ASSOCIATION. 2006. *KNX Journal 2/2006*. Periodical.

[Kubitschke *et al.*, 2009] KUBITSCHKE, L., GAREIS, K., LULL, F., MUELLER, S., CULLEN, K., DELANEY, S., TAYLOR, L. Q., WYNNE, R., & RAUHALA, M. 2009 (January). *ICT & Ageing: Users, Markets and Technologies | Compilation Report on Ethical Issues*. Online. http://www.ict-ageing.eu/ict-ageing-website/wp-content/uploads/2008/11/d11_ethics_compilation_rep_with_exec_sum.pdf (as of 14 Aug 2009).

[Kumpch *et al.*, 2010] KUMPCH, M., LUIZ, T., & MADLER, C. 2010. Analyse der Einsatzdaten eines innerklinischen Notfallteams. *Anaesthesist*, **59**, 217–224.

[Kung *et al.*, 2007] KUNG, H. Y., HSU, C. Y., & LIN, M. H. 2007. Sensor-based pervasive healthcare system: Design and implementation. *Journal Of High Speed Networks*, **16**(1), 35–49.

[Le *et al.*, 2007] LE, ANKANG, JAITNER, THOMAS, & LITZ, LOTHAR. 2007. Sensor-based Training Optimization of a Cyclist Group. *Pages 265–270 of: Proceedings of HIS2007 (7th International Conference on Hybrid Intelligent Systems)*.

[Lee, 2008] LEE, E. A. 2008 (May 5–7,). Cyber Physical Systems: Design Challenges. *Pages 363–369 of: Proc. 11th IEEE International Symposium on Object Oriented Real-Time Distributed Computing (ISORC)*.

[Leijdekkers *et al.*, 2007] LEIJDEKKERS, PETER, GAY, VALERIE, & LAWRENCE, ELAINE. 2007 (Jan.). Smart Homecare System for Health Tele-monitoring. *Pages 3–3 of: Digital Society, 2007. ICDS '07. First International Conference on the*.

[Lewis & Rieman, 1993-1994] LEWIS, C., & RIEMAN, J. 1993-1994. *Task-centered User Interface Design | A Practical Introduction*. Online. http://hcibib.org/tcuid/tcuid.pdf (as of 22 Aug 2009).

[Lindwer *et al.*, 2003] LINDWER, M., MARCULESCU, D., BASTEN, T., ZIMMENNANN, R., MARCULESCU, R., JUNG, S., & CANTATORE, E. 2003. Ambient intelligence visions and achievements: linking abstract ideas to real-world concepts. *Pages 10–15 of: Proc. Design, Automation and Test in Europe Conference and Exhibition*.

[Litz, 2005] LITZ, LOTHAR. 2005. *Grundlagen der Automatisierungstechnik*. Oldenbourg Wissensch.Vlg.

[Luiz *et al.*, 2000] LUIZ, TH., HUBER, TH., SCHIETH, B., & MADLER, C. 2000. Einsatzrealität eines städtischen Notarztdienstes: Medizinisches Spektrum und lokale Einsatzverteilung. *Anaesthesiologie und Intensivmedizin*, **10**, 765–773.

[Lukowicz *et al.*, 2004] LUKOWICZ, P., KIRSTEIN, T., & TROSTER, G. 2004. Wearable systems for health care applications. *Methods Of Information In Medicine*, **43**(3), 232–238.

[Lunze, 2007] LUNZE, JAN. 2007. *Automatisierungstechnik*. Oldenbourg Wissensch.Vlg.

[Lyu, 1996] LYU, MICHAEL R. 1996. *Handbook of software reliability engineering*. Los Alamitos, Calif.: IEEE Computer Society Press et al.

[Mahoney *et al.*, 2007] MAHONEY, D. F., PURTILO, R. B., WEBBE, F. M., ALWAN, M., BHARUCHA, A. J., ADLAM, T. D., JIMISON, H. B., TURNER, B., & BECKER, S. A. 2007. In-home monitoring of persons with dementia: Ethical guidelines for technology research and development. *Alzheimers Dement*, **3**(3), 217–226.

[Manzeschke, 2009] MANZESCHKE, A. 2009. Ethische Implikationen des Ambient Assisted Living - ein Problemaufriss. *In: Proceedings 2. Deutscher AAL-Kongress*. Document ID: 12.1.

[Matthews *et al.*, 2006] MATTHEWS, ZOE, CHANNON, ANDREW, & VAN LERBERGHE, WIM. 2006. *Will there be enough people to care? Notes on workforce implications of demographic change 2005-2050*. Tech. rept. World Health Organization.

[May & Zimmer, 1996] MAY, ELAINE L., & ZIMMER, BARBARA A. 1996. The Evolutionary Development Model for Software. *hp journal*, **48**, 39–45.

[McCrie, 1988] MCCRIE, ROBERT D. 1988. The Development of the U. S. Security Industry. *Annals of the American Academy of Political and Social Science*, **498**, 23–33.

[McGill *et al.*, 1978] MCGILL, ROBERT, TUKEY, JOHN W., & LARSEN, WAYNE A. 1978. Variations of Box Plots. *The American Statistician*, **32**(1), 12–16.

[Mira, 2008] MIRA, JOSÉ MIRA. 2008. Symbols versus connections: 50 years of artificial intelligence. *Neuro-comput.*, **71**(4-6), 671–680.

[Munakata, 2008]MUNAKATA, TOSHINORI. 2008. *Fundamentals of the New Artificial Intelligence: Neural, Evolutionary, Fuzzy and More (Texts in Computer Science)*. Springer Publishing Company, Incorporated.

[Mynatt *et al.*, 2000] MYNATT, ELIZABETH D., ESSA, IRFAN, & ROGERS, WENDY. 2000. Increasing the opportunities for aging in place. *Pages 65–71 of: CUU '00: Proceedings on the 2000 conference on Universal Usability.* New York, NY, USA: ACM.

[Nambu *et al.*, 2000] NAMBU, M., NAKAJIMA, K., KAWARADA, A., & TAMURA, T. 2000 (9-10 Nov.). The automatic health monitoring system for home health care. *Pages 79–82 of: Information Technology Applications in Biomedicine, 2000. Proceedings. 2000 IEEE EMBS International Conference on.*

[Nehmer, 2009] NEHMER, J. 2009. Elektronische Notfallüberwachung für Alleinlebende. *Notfall & Rettungsmedizin*, **12**, 19–24.

[Nehmer *et al.*, 2006] NEHMER, JÜRGEN, BECKER, MARTIN, KARSHMER, ARTHUR, & LAMM, ROSEMARIE. 2006. Living assistance systems: an ambient intelligence approach. *Pages 43–50 of: ICSE '06: Proceeding of the 28th international conference on Software engineering.* New York, NY, USA: ACM Press.

[Noury *et al.*, 2000]NOURY, N., HERVE, T., RIALLE, V., VIRONE, G., MERCIER, E., MOREY, G., MORO, A., & PORCHERON, T. 2000 (Oct. 12–14,). Monitoring behavior in home using a smart fall sensor and position sensors. *Pages 607–610 of: Proc. Conference On Microtechnologies in Medicine and Biology, 1st Annual International 2000.*

[NSF, 2006] NSF, NATIONAL SCIENCE FOUNDATION. 2006. *NSF Workshop On Cyber-Physical Systems - Research Motivation, Techniques and Roadmap.* Online. http://varma.ece.cmu.edu/CPS/ (as of 12 Nov 2009).

[Ohashi *et al.*, 2004] OHASHI, K, BLEIJENBERG, G, VAN DER WERF, S, PRINS, J, AMARAL, LAN, NATELSON, BH, & YAMAMOTO, Y. 2004. Decreased fractal correlation in diurnal physical activity in chronic fatigue syndrome. *METHODS OF INFORMATION IN MEDICINE*, **43**(1), 26–29. 4th International Workshop on Biosignal Interpretation (BSI2002), Villa Olmo, ITALY, JUN 24-26, 2002.

[Paulus & Romanowski, 2009] PAULUS, WOLFGANG, & ROMANOWSKI, SASCHA. 2009. *Telemedizin und AAL in Deutschland: Geschichte, Stand und Perspektiven.* Tech. rept. 9. Institut Arbeit und Technik.

[Peirce, 1852] PEIRCE, BENJAMIN. 1852. Criterion for the Rejection of Doubtful Observations. *The Astronomical Journal*, **2**(21), 161–163.

[Reyes Alamo *et al.*, 2009] REYES ALAMO, JOSE M., WONG, JOHNNY, BABBITT, RYAN, YANG, HEN-I, & CHANG, CARL K. 2009. Using Web Services for Medication Management in a Smart Home Environment. *Pages 265–268 of: ICOST '09: Proceedings of the 7th International Conference on Smart Homes and Health Telematics.* Berlin, Heidelberg: Springer-Verlag.

[Rombach & Ulery, 1989] ROMBACH, H.D., & ULERY, B.T. 1989. Improving software maintenance through measurement. *Proceedings of the IEEE*, **77**(4), 581 –595.

[Roscoe *et al.*, 2007] ROSCOE, J.A., KAUFMAN, M.E., MATTESON-RUSBY, S.E., PALESH, O.G., RYAN, J.L., KOHLI, S., PERLIS, M.L., & MORROW, G.R. 2007. Cancer-Related Fatigue and Sleep Disorders. *Oncologist*, **12**(suppl_1), 35–42.

[Rose, 1995] ROSE, MICHAEL. 1995. *Gebäudesystemtechnik in Wohn- und Zweckbau mit dem EIB.* 2., bearb. aufl. edn. Heidelberg: Huethig.

[Royce, 1987] ROYCE, W. W. 1987. Managing the development of large software systems: concepts and techniques. *Pages 328–338 of: Proceedings of the 9th international conference on Software Engineering.* ICSE '87. Los Alamitos, CA, USA: IEEE Computer Society Press.

[Schulze et al., 2009] SCHULZE, BERND, FLOECK, MARTIN, & LITZ, LOTHAR. 2009. Concept and Design of a Video Monitoring System for Activity Recognition and Fall Detection. *Pages 182–189 of: Proceedings of ICOST (International Conference on Smart Homes and Health Telematics).*

[Shaji et al., 2009] SHAJI, S, BOSE, SRIJA, & KURIAKOSE, SHAN. 2009. Behavioral and psychological symptoms of dementia: A study of symptomatology. *Indian Journal of Psychiatry*, **51**, 38–41.

[Sixsmith, 2009] SIXSMITH, ANDREW. 2009. Understanding the Older User of Ambient Technologies. *Pages 511–519 of: Proceedings of the 13th International Conference on Human-Computer Interaction. Part III.* Berlin, Heidelberg: Springer-Verlag.

[Smith & House, 1969] SMITH, RANDOLF J., & HOUSE, KENNETH R. 1969. *United States Patent No. 3,460,124: Smoke Detector.*

[Spellerberg et al., 2009] SPELLERBERG, ANNETTE, GRAUEL, JONAS, & SCHELISCH, LYNN. 2009. Ambient Assisted Living - Ein erster Schritt in Richtung eines technisch-sozialen Assistenzsystems für ältere Menschen. *Hallesche Beiträge zu den Gesundheits- und Pflegewissenschaften*, **8**(39).

[Storf et al., 2009] STORF, HOLGER, KLEINBERGER, THOMAS, BECKER, MARTIN, SCHMITT, MARIO, BOMARIUS, FRANK, & PRUECKNER, STEPHAN. 2009. An Event-Driven Approach to Activity Recognition in Ambient Assisted Living. *Pages 123–132 of: Proceedings of AmI 2009.*

[Taylor, 1996] TAYLOR, JOHN R. 1996. *An Introduction to Error Analysis: The Study of Uncertainties in Physical Measurements.* 2 sub edn. University Science Books.

[Tognazzini, 1992] TOGNAZZINI, BRUCE. 1992. *Tog on Interface.* Addison-Wesley Publishing.

[Tukey, 1990] TUKEY, JOHN W. 1990. Data-Based Graphics: Visual Display in the Decades to Come. *Statistical Science*, **5**(3), 327–339.

[Turing, 1950] TURING, ALAN M. 1950. Computing Machinery and Intelligence. *Mind*, **59**, 433–460.

[van der Poel et al., 2004] VAN DER POEL, C., PESSOLANO, F., ROOVERS, R., WIDDERSHOVEN, F., VAN DE WALLE, G., AARTS, E., & CHRISTIE, P. 2004 (Sept. 21–23,). On ambient intelligence, needful things and process technologies. *Pages 3–10 of: Proc. Proceeding of the 34th European Solid-State Device Research conference ESSDERC 2004.*

[VDIVDE, 2007] VDIVDE. 2007. *VDI/VDE 2180: Sicherung von Anlagen der Verfahrenstechnik mit Mitteln der Prozessleittechnik (PLT) - Blatt 3: Anlagenplanung, -errichtung und -betrieb.*

[Virone & Sixsmith, 2008] VIRONE, G., & SIXSMITH, A. 2008. Monitoring activity patterns and trends of older adults. *Conf Proc IEEE Eng Med Biol Soc*, **2008**, 2071–2074.

[von Foerster, 2008] VON FOERSTER, MICHAEL. 2008 (June). *Ambient Assisted Living.* ZVEI Press Conference Transcript.

[Wang, 2007] WANG, PEI. 2007. Three fundamental misconceptions of Artificial Intelligence. *J. Exp. Theor. Artif. Intell.*, **19**(3), 249–268.

[Weiser, 1991] WEISER, MARK. 1991. The computer for the 21st century. *Scientific American*, **265**(3), 66–75.

[Winters & Wang, 2003] WINTERS, J. M., & WANG, YU. 2003. Wearable sensors and telerehabilitation. *IEEE Engineering in Medicine and Biology Magazine*, **22**(3), 56–65.

[Wolf, 2009] WOLF, W. 2009. Cyber-physical Systems. *Computer*, **42**(3), 88–89.

[Wong, 1984] WONG, CAROLYN. 1984. A Successful Software Development. *IEEE Trans. Software Eng.*,
        10(6), 714–727.

[Yu *et al.*, 2009] YU, XINGUO, WANG, XIAO, NGAM, PANACHIT K., ENG, HOW L., & CHEONG, LOONG F. 2009.
        Fall Detection and Alert for Ageing-at-Home of Elderly. *Pages 209–216 of: ICOST '09: Proceedings
        of the 7th International Conference on Smart Homes and Health Telematics.* Berlin, Heidelberg:
        Springer-Verlag.

[Zhang *et al.*, 2006] ZHANG, TONG, WANG, JUE, LIU, PING, & HOU, JING. 2006. Fall Detection by Embedding an
        Accelerometer in Cellphone and Using KFD Algorithm. *International Journal of Computer Science
        and Network Security*, **6**(10), 277–284.

[Zito *et al.*, 2007] ZITO, D., PEPE, D., NERI, B., DE ROSSI, D., LANATA, A., TOGNETTI, A., & SCILINGO, E. P. 2007
        (Aug. 22–26,). Wearable System-on-a-Chip UWB Radar for Health Care and its Application to the
        Safety Improvement of Emergency Operators. *Pages 2651–2654 of: Proc. 29th Annual International
        Conference of the IEEE Engineering in Medicine and Biology Society EMBS 2007.*

10.2.   Index of Tables

10.3. Index of Figures

10.4.   Frequently Used Symbols and Abbreviations

α	smoothing factor
AAL	Ambient Assistive Living
ADI_p	average duration of inactivity while tenant present
ADL	activities of daily living
AI	Artificial Intelligence
AmI	Ambient Intelligence
ALT	alarm threshold
AT	assistive technology
CFS	Chronic Fatigue Syndrome
CPS	Cyber Physical Systems
DI	elapsed time after last observed activity (duration of inactivity)
DI_p	elapsed time after last observed activity while tenant present
DIE_p	duration of inactivity without outliers while tenant present
DIE_p^*	smoothed duration of inactivity without outliers while tenant present
EIB	European Installation Bus (Europäischer Installationsbus)
EMS	emergency medical service
FSM	finite state machine
GUI	graphical user interface
HCA	human-centred approach
HVAC	heating, ventilation, and air conditioning
ICT	information and communication technologies
KNX	Konnex, successor of EIB (EN 50090, ISO/IEC 14543)
LB	lower boundary
MDI	maximum duration of inactivity
MDI_p	maximum duration of inactivity while tenant present
MIT	multi-day inactivity-derived threshold
MTFA	mean time between false alarms
mSingSLAT	masked singular static linear alarm threshold
mSLAT	masked static linear alarm threshold
OMIP	outlier-free, multi-day inactivity pattern
ORC	occupied room criterion
P(t, d)	number of days out of d days on which tenant was present at time t
PERS	personal emergency response system (medical alarm)
PO	period of operation
pp	percentage point (arithmetic difference of two percentages)
S_n	set of n days
SLAT	static linear alarm threshold
UB	upper boundary
UC	ubiquitous computing